T0220293

Physics: From Classical to Quantum

(A Self-Teaching Guide)

Joseph Palazzo

authorHOUSE

AuthorHouse™
1663 Liberty Drive
Bloomington, IN 47403
www.authorhouse.com
Phone: 1 (800) 839-8640

Published by AuthorHouse 02/12/2020

ISBN: 978-1-7283-4674-8 (sc)
ISBN: 978-1-7283-4673-1 (e)

Library of Congress Control Number: 2020902663

Print information available on the last page.

To Xander and Théo

Preface

This book is not for everyone but for those who want to learn physics and are willing to make that commitment. If you fulfill that profile then this book will be of great help. It is unique in the sense that it takes the reader from Classical Physics to Quantum Mechanics and on to Quantum Field Theory. Physics belongs not only to the professionals but to everyone and anyone who so desire to learn and acquire a fundamental understanding. So part of the aim of this book is to reach other scientists, engineers, members of the medical field, and members of the community at large who want more than what is found on websites, science magazines and popular books. Moreover there is an urgent need to understand physics at its most basic core – that is, in the language of mathematics. It is in this spirit that this book was written.

On a technical level, the aim of this book is to bring the reader on a voyage from Classical physics to Quantum Field Theory in the **shortest possible path**. The purpose is twofold: for the student in a physics program who wants to learn physics at an accelerating pace within a very short period; but also for the individual who wants to learn physics at a more leisure pace over a longer period of time. This book is a lifetime investment. At a little over 400 pages, it is compact such that one can bring it anywhere at any time.

Prerequisite: high school physics. A rudimentary knowledge of calculus would help. But the appendices covering most of the math, calculus included, are sufficient to carry the reader through the minimum path presented in this book. As much as it is possible to do, there are as many steps spelled out in detail so that the reader does not get bogged down by the mathematics, which often can spoil the joy of following a step-by-step derivation. The reader is invited to review the appendices as often as necessary. They cover the basics in calculus (derivatives and integrals), linear algebra (vectors and matrices), group theory and miscellaneous special functions and integrals. This book intends to demystify, and not mystify. The assumption is that the reader is curious and willing to embark on a significant intellectual challenge.

Chapter 1 introduces the reader to the concepts that are ubiquitous in Newtonian physics – position, momentum, energy and fields. Linear, rotational and wave motions are described by their pertinent equations. Chapter 2 deals with the Lagrangian and Hamiltonian formalism both for particles and fields via the Principle of Least Action. The chapter ends with the important Noether theorem which relates the concept of symmetry with the concept of conservation laws. Chapter 3 reviews the Maxwell equations for electric and magnetic fields. And also what's most pertinent to fields: gauge invariance. This is illustrated with the electromagnetic field tensor. Chapter 4 is concerned with the basic concepts of thermodynamics such as volume, pressure, temperature and entropy. Chapter 5 describes the essential ideas of Special Relativity in regard to the space-time interval, the Lorentz transformation and Lorentz invariance principle. A derivation of Einstein famous equation $E=mc^2$ is presented. Chapter 6 gives an outline of some of the key ideas in General Relativity – the Equivalence Principle, the Riemann Curvature of Space, the Einstein's Field equations, and the Schwarzschild's radius. In part 2 chapter 7 the reader is introduced to the fundamentally different approach of Quantum Mechanics. Here we separate the notion of a state from its observable so that position, momentum and energy (fields will be taken up in part 3) which are observables are now represented as operators in the theory. The quantum states are found in the wave function which in essence is used to calculate probabilities. Also QM requires that real-valued observables be represented by Hermitian operators, and operators acting on the wave function must be unitary in order that the sum of all outcomes equals one, a fundamental mathematical principle of Probability theory. Chapter 8 on Quantum Statistical Mechanics is a bird's eye view on bosons, designated by commutation relationship, and fermions, designated by anti-commutation relationship, both particles obeying different statistical distributions. This sets the stage for Planck's famous insights into black body radiation. Chapter 9 presents the ideas behind quantum tunneling, emphasizing the difference between Classical physics and Quantum Mechanics. And chapter 10 develops two important equations in Relativistic Quantum Mechanics: the Klein-Gordon equation and the Dirac equation. Chapter 11 in part 3 moves the reader into the concept of fields. The essential idea is that the wave function metamorphoses into a field operator, and quantum states are now expressed by the annihilation/creation operators acting on the vacuum state. The emphasis is on the Canonical formalism. We conclude the chapter by showing that QFT is Lorentz invariant in order to be compatible with Special Relativity. Chapter 12 deals with the core concepts of QFT, that is, the time-evolution operator, propagators, the S-matrix, and Feynman's rules for interactions. Chapter 13 takes a brief view of the Path Integral Formalism as an alternative to calculating probability amplitudes by adding contributions from all possible trajectories. Chapter 14 reprises the concept of gauge theory and how one can proceed to the Higgs Mechanism - a procedure that gives mass to a massless scalar particle while at the same time keeping the theory Lorentz invariant (compatible with SR) and unitary (QM is a probability theory).

There are many topics which were left out but necessary in order to keep the size of this book at a reasonable level. Nevertheless the core ideas which are covered will give the reader a taste as to why physics is the crown jewel of human achievements.

Some historical factoids: these dates were chosen on the grounds of some specific publications. It certainly doesn't mean that no contribution or advancement was accomplished after these dates.

(1) Classical physics was developed circa 1632[1] (Galileo's Law of Inertia) to 1900[2], which marks the publication of Planck's paper on Quantum Mechanics (QM).

(2) QM: from 1900 to 1927[3], the year of Dirac's paper, considered to be the first on Quantum Field Theory (QFT).

(3) QFT: from 1927 to 1971[4], the year of Wilson's paper, the final contribution on the Renormalization Group.

If we lump QM and QFT together as Quantum Physics (QP), that era spans some 70 years. Since then the foundation of QP has been consolidated, written in a notational convention and simplified presentation, making it more accessible to a much larger segment of the general public. This is another aim of this book: to fill the gap and provide easier accessibility on a subject notoriously known as difficult.

[1] American Journal of Physics 32, 601 (1964)

[2] Planck M 1900 Zur Theorie des Gesetzes der Energieverteilung im Normalspectrum Verhandlungen der Deutschen Physikalischen Gesellschaft 2, 237–45

[3] Dirac, P.A.M. 1927, Quantum Theory of emission and absorption of radiation, Proc. Roy. Soc. London A, 114, 243-265.

[4] Wilson K.G. 1971, Renormalization group and critical phenomena 1. Renormalization and the Kadanoff scaling picture. Phys.Rev. B, 4, 3174-3183.

Content

Part 1

CLASSICAL PHYSICS

Chapter 1

Newtonian Formalism

The fundamental problem of physics is to determine the trajectory of a particle as it moves through space. It is this motion that allows us to define the present (where it is right now), the past (where was it before) and the future (where will it be) so that our next task is to find the equation of motion (eom) of that particle. To assist us in this task, we introduce a coordinate system, denoted by (x,y,z) in a three-dimensional (3-D) world. But in many situations, one can always reduce the problem to a 1-D system. For instance, an object is moving along a coordinate system such as in Fig. 1A.1 below. But nothing stops us from choosing a coordinate system along the path the object is following, thus reducing the study to a 1-D problem. Whatever laws govern along that path will also apply long the x-direction, the y-direction, and the z-direction. This is the concept of linearity in its most raw form: each component along each axis acts independently of each other. We can combine these results by simply adding them as vectors. A lot of what we will be doing is just that: how to deal with vectors and the algebra that accompanies them (see appendix D for more on linearity and vectors).

1A. Kinematics

In a given reference frame, the position of a particle can be specified by a vector. In Cartesian coordinates, the position vector is,

(1A.1a) $r = ix + jy + kz$

Where $i, j,$ and k are unit vectors along the three axes labelled x, y and z, respectively. And x, y and z are the components of the vector. Note: **vectors in bold-face font.**

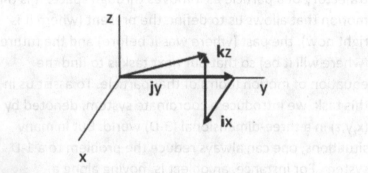

Fig. 1A.1

(1A.1b) The magnitude of r is designated as $|r|$, or r.

$$\rightarrow r^2 = (ix + jy + kz) \cdot (ix + jy + kz)$$

$$= x^2 + y^2 + z^2 \text{ (see Appendix D, equation DA.3a)}$$

Or $r = (x^2 + y^2 + z^2)^{½}$

The velocity is the derivative (see appendix E, equation EA.1 for the definition of a derivative) of the position vector with respect to the time.

(1A.2) $v \equiv \dfrac{dr}{dt} \equiv \dot{r} \equiv i\dot{x} + j\dot{y} + k\dot{z}$

Where the "d" stands for derivative, and dots indicate differentiation with respect to t.

The speed is the magnitude of the velocity, and it is given as,

(1A.3) $v = |\mathbf{v}| = (\dot{x}^2 + \dot{y}^2 + \dot{z}^2)^{1/2}$

The time derivative of the velocity is the acceleration,

(1A.4) $\mathbf{a} \equiv \dfrac{d\mathbf{v}}{dt} \equiv \dfrac{d\dot{\mathbf{r}}}{dt} \equiv \dfrac{d^2\mathbf{r}}{dt^2} \equiv \ddot{\mathbf{r}}$

In Cartesian coordinates we have,

(1A.5) $\mathbf{a} = \mathbf{i}\ddot{x} + \mathbf{j}\ddot{y} + \mathbf{k}\ddot{z}$

Some useful, recurring formulas (in 1-D):

From equation 1A.2, with $(\dot{y} = 0, \dot{z} = 0)$

(1A.6) $dx = vdt$

(1A.7) For v as a constant, integrate both sides, (see appendix E, equation EB1 for the definition of an integral)

$\rightarrow \int_{x_o}^{x_f} dx = \int_{t_o}^{t_f} vdt$

$\rightarrow x_f - x_o = vt_f - vt_o$

(1A.8) For a body starting from rest ($v(t_o) = 0$) at x_o,

$\rightarrow x_f = x_o + vt_f$

Note: this is an equation of a straight line for a body moving at constant speed. Such a particle is said to be free (no net force acting on it, section 1C).

(1A.9a) For the acceleration a as a constant, integrate the first part of equation 1A.4,

$\displaystyle\int_{v_o}^{v} dv = \int_{t_o}^{t_f} adt$

$\rightarrow v - v_0 = at_f - at_o,$

25

(1A.9b) Or, $v = v_0 + a\Delta t$

Where $\Delta t = t_f - t_o$

(1A.10a) For a body starting at $t_o = 0$

$\int_{x_o}^{x_f} dx = \int_0^t v dt'$ (equation 1A.7)

Substitute equation 1A.9b,

$\rightarrow x_f - x_o = \int_0^t (v_0 + at') dt'$

$$= v_0 t + \int_0^t at' \, dt' = v_0 t + \tfrac{1}{2}at^2$$

(1A.10b) Or, $x_f = x_o + v_0 t + \tfrac{1}{2}at^2$

Example: projectiles are example of what we mentioned about vectors - each component acts independently.

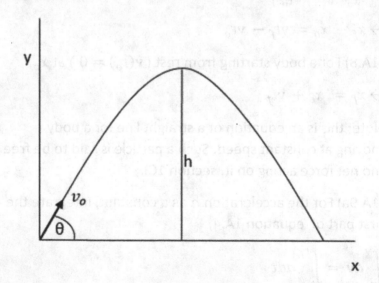

Fig. 1A.2

Consider the velocity components:

(1A.11a) In the x-direction: $v_x = v_o cos\theta$

(1A.11b) In the y-direction: $v_y = v_o sin\theta - gt$

Where g is the acceleration due to gravity, acting downward. The displacement, $x_o = 0$:

(1A.11c) Along the x-axis x = $v_o t cos\theta$ (from 1A.8)

(1A.11d)

Along the y-axis: y = $v_o t sin\theta - \frac{1}{2}gt^2$ (from 1A.10b)

(1A.12) The maximum height (h) is reached when $v_y = 0$

$\rightarrow 0 = v_o sin\theta - gt_h$ (equation 1A.11b)

$\rightarrow t_h = \frac{v_o sin\theta}{g}$

(1A.13a) So we can calculate the height h:

$h = v_o t_h sin\theta - \frac{1}{2}gt_h^2$ (equation 1A.11d)

Substitute equation 1A.12 for t_h,

$$= v_o \frac{v_o sin\theta}{g} sin\theta - \frac{1}{2}g \left(\frac{v_o sin\theta}{g}\right)^2 = \frac{v_o^2 sin^2\theta}{2g}$$

(1A.13b) Or, $g = \frac{v_o^2 sin^2\theta}{2h}$

The total time of flight is $2t_h$, (going up plus going down) therefore the total displacement (range) along the x-axis:

(1A.14) $R = 2v_o t_h cos\theta$ (1A.11c)

$$= 2v_o \frac{v_o sin\theta}{g} cos\theta$$ (from 1A.12)

Using the previous result for h, we can eliminate g, (equation 1A.13b):

$$R = 2v_o \frac{v_o sin\theta}{\frac{v_o^2 sin^2\theta}{2h}} cos\theta = 4h\, cot\theta$$

In its standard form: $h = \frac{R}{4} tan\theta$ (using equation CE.4)

Example: circular motion

Consider a body tied to a rope and forced to move in a circle with radius R (Fig. 1A.3A). We see that its change in position is perpendicular to the radius of that circle.

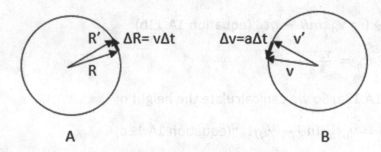

A   B

Fig. 1A.3

The period T is defined as the time taken to go through one full rotation. So the magnitude of its velocity is then,

(1A.15) $v = \frac{2\pi R}{T}$ (1A.8)

Similarly, the change in velocity is perpendicular to the velocity (Fig. 1A.3B). The magnitude of the acceleration is,

(1A.16a) $a = \frac{2\pi v}{T}$ (1A.9)

Substitute for the velocity we get,

28

(1A.16b) $a = \dfrac{2\pi}{T}\dfrac{2\pi R}{T} = \dfrac{4\pi^2 R}{T^2}$

(1A.17a) If we divide the two equations, we get

(1A.17b) $\dfrac{a}{v} = \dfrac{2\pi v/T}{2\pi R/T} = \dfrac{v}{R}$

Or $a = \dfrac{v^2}{R}$

1B. Polar coordinates

As we shall see, there are additional features when we use polar coordinates. But first we define the polar coordinates as,

(1B.1) $x = r\cos\theta;\ y = \sin\theta$

Now the position vector can be represented by a radial distance $r = |r|$ and a unit radial vector λ_r (Fig. 1B.1).

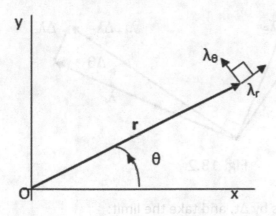

Fig. 1B.1

Note that λ_θ, the unit transverse vector, is perpendicular to λ_r. The position vector can be expressed as,

29

(1B.2) $r = r\lambda_r$

Again we take the derivative to get the velocity,

(1B.3) $v = \dfrac{dr}{dt} = \dfrac{d(r\lambda_r)}{dt} = \dot{r}\lambda_r + r\dfrac{d\lambda_r}{dt}$

We get an extra term and now we must determine what that second term is. Consider Fig. 1B.2, where the position vector has moved by an angle $\Delta\theta$. Both unit vectors λ_r and λ_θ, which are perpendicular to each other, are rotated by the same angle.

As $\Delta\theta \to 0$ we have $\Delta\lambda_r$ parallel to λ_θ, and so we can say,

(1B.4) $\Delta\lambda_r \approx \lambda_\theta\,\Delta\theta$

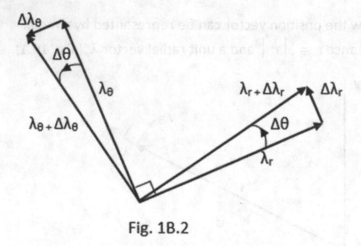

Fig. 1B.2

Divide both sides by Δt, and take the limit:

(1B.5) $\dfrac{d\lambda_r}{dt} = \lambda_\theta\dfrac{d\theta}{dt} = \lambda_\theta\dot{\theta}$

Where $\dot{\theta}$ is the angular velocity.

Similarly, we have $\Delta\lambda_\theta$ anti-parallel to λ_r, and so

(1B.6) $\Delta\lambda_\theta \approx -\lambda_r \Delta\theta$

And therefore,

(1B.7) $\frac{d\lambda_\theta}{dt} = -\lambda_r \frac{d\theta}{dt} = -\lambda_r \dot\theta$

Substitute equation 1B.5 into 1B.3,

(1B.8) $v = \dot r \lambda_r + r \frac{d\lambda_r}{dt} = \dot r \lambda_r + r\dot\theta\lambda_\theta$

Note that we started with just a radial position vector, but now the velocity vector has two components, a radial and a transverse.

To get the acceleration, we differentiate equation 1B.8 with respect to t,

(1B.9) $a = \frac{dv}{dt} = \ddot r \lambda_r + \dot r \frac{d\lambda_r}{dt} + \frac{d(r\dot\theta)}{dt}\lambda_\theta + r\dot\theta \frac{d\lambda_\theta}{dt}$

Substitute equation 1B.5 into the second term, and equation 1B.7 into the last term,

(1B.10) $a = \ddot r \lambda_r + \dot r \dot\theta \lambda_\theta + \frac{d(r\dot\theta)}{dt}\lambda_\theta - r\dot\theta^2\lambda_r$

$\qquad = \ddot r \lambda_r + \dot r \dot\theta\lambda_\theta + (\dot r\dot\theta + r\ddot\theta)\lambda_\theta - r\dot\theta^2\lambda_r$

$\qquad = (\ddot r - r\dot\theta^2)\lambda_r + (2\dot r\dot\theta + r\ddot\theta)\lambda_\theta$

Where in the 2nd line we used the Leibnitz rule (see appendix E, equation EA.9b) on the 3rd term. In the 3rd line, the acceleration has a transverse component and a negative term $(-r\dot\theta^2)$ pointing inward along the radial component.

1C. Dynamics

As stated at the beginning of this chapter, our main concern is to determine the position of a particle on a trajectory. Part of this problem is to find out which factors are relevant in determining that future course. We will now look at how the notion of force comes into this picture. We start with Newton's three laws of motion.

The first law is also known as Galileo's Law of Inertia.

It used to be believed that in order to move, you needed a force. This was based on everyday observations: a carriage had to be pulled by a horse; you needed to row a boat or have a sail nailed to the boat and let the wind push the boat. This notion which dominated the Western world went all the way to Aristotle for nearly two thousand years. So the experiment that would overturn this idea is perhaps the most important one, at least as far as science is concerned. Galileo discovered one instance in which a body could be in motion without the need of a force – and these are Galileo's famous inclined plane experiments.

Consider how Galileo arrived at the Law of Inertia.

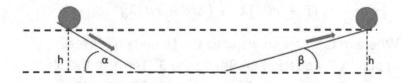

Fig. 1C.1

A ball is released from a height h on an inclined plane making an angle α with the ground (Fig. 1C.1). It would then roll down and climb a second inclined plane making an angle β with the ground. What Galileo observed is that the ball would travel far enough until it would climb the same height h where it would come to rest temporarily and then reverse course. By varying the angle of the second plane, he always observed that the ball would climb to a height h before reversing direction. See Fig. 1C.2.

Fig. 1C.2

This led him to a thought experiment: what would happen if the second inclined plane were to be removed?

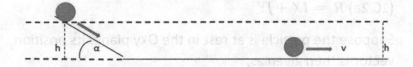

Fig. 1C.3

He reasoned that the ball would continue endlessly in a straight line with a velocity v trying to reach this height h, unless other forces would compel it to change its velocity (Fig. 1C.3). Now this doesn't happen in the real world – and so Galileo further deduced that there are other forces acting on the ball, namely friction that would bring it to rest.

And it was Newton who seized upon Galileo's insights and formulated the three laws of motion. We now come to his second law of motion, which reads as:

(1C.1) $F = ma = m \frac{dv}{dt} = m\ddot{r}$

Where F means a <u>net</u> force acting on a body. In the MKS system, a one kilogram mass accelerated at one meter per second per second is acted upon by a net force of one newton. In terms of unit dimension: $[F] = [\frac{ML}{T^2}]$(see appendix B for units).

Example: Consider a coordinate system Oxy rotating with angular speed ω in a plane of an inertial system OXY which is at rest, as shown in Fig. 1C.4. Note that we take the origin to be coincident without loss of generality. In the OXY plane, the position vector of a particle is written as,

(1C.2a) $R = IX + JY$

Suppose the particle is at rest in the Oxy plane. Its position vector is then given as,

(1C.2b) $r = ix + jy$

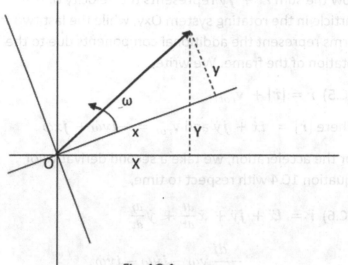

Fig. 1C.4

Now suppose that the particle is no longer at rest in the Oxy plane but instead moving in that frame. For the velocity we take the derivative with respect to the time, taking into account that the unit vectors i and j are also moving.

(1C.2c) $\dfrac{dr}{dt} = i\dot{x} + x\dfrac{di}{dt} + j\dot{y} + y\dfrac{dj}{dt}$

From equations 1B.5 and 1B.7, we have,

(1C.3) $\dfrac{di}{dt} = j\dfrac{d\theta}{dt} = j\omega$ and $\dfrac{dj}{dt} = -i\dfrac{d\theta}{dt} = -i\omega$

Where $\dfrac{d\theta}{dt} = \dot{\theta} = \omega$ (equation 1B.5)

Substituting,

(1C.4) $\dot{r} = i\dot{x} + j\dot{y} - iy\omega + jx\omega$

Now the sum $i\dot{x} + j\dot{y}$ represents the velocity of the particle in the rotating system Oxy, while the last two terms represent the additional components due to the rotation of the frame. We write,

(1C.5) $\dot{r} = [\dot{r}] + \mathbf{v}_{rot}$

Where $[\dot{r}] = i\dot{x} + j\dot{y}$ and $\mathbf{v}_{rot} = -iy\omega + jx\omega$

For the acceleration, we take a second derivative of equation 1C.4 with respect to time,

(1C.6) $\ddot{r} = i\ddot{x} + j\ddot{y} + \dot{x}\dfrac{di}{dt} + \dot{y}\dfrac{dj}{dt}$

$$-\dfrac{di}{dt}y\omega - i\dot{y}\omega - iy\dot{\omega}$$

$$+\dfrac{dj}{dt}x\omega + j\dot{x}\omega + jx\dot{\omega}$$

$$= i\ddot{x} + j\ddot{y} + \dot{x}j\omega - \dot{y}i\omega$$

$$-j\omega y\omega - i\dot{y}\omega - iy\dot{\omega}$$

$$-i\omega x\omega + j\dot{x}\omega + jx\dot{\omega}$$

Where we substitute equation 1C.3 in the second equation. Regrouping,

(1C.7) $\ddot{r} = i\ddot{x} + j\ddot{y} - 2\omega(i\dot{y} - j\dot{x})$

$$-\omega^2(ix + jy) - \dot{\omega}(iy - jx)$$

Again we recognize that the sum $i\ddot{x} + j\ddot{y}$ is the acceleration in the rotating system, which we denote as $[\ddot{r}]$.

Rearranging we have

(1C.8)

$$[\ddot{r}] = \ddot{r} + 2\omega(i\dot{y} - j\dot{x}) + \omega^2(ix + jy) + \dot{\omega}(iy - jx)$$

To get the force associated with the acceleration we multiply by the mass m,

(1C.9) $m[\ddot{r}] = F + F_{Cor} + F_{cent} + F_{trans}$

Where

$$F = m\ddot{r}$$

This is the force experienced in the fixed frame. But included are the fictitious forces, namely:

$$F_{Cor} = 2m\omega(i\dot{y} - j\dot{x})$$

Which is the Coriolis force.

$$F_{cent} = m\omega^2(ix + jy) = m\omega^2 r$$

This is the centrifugal force. And lastly, the transverse force is,

$$F_{trans} = m\dot{\omega}(iy - jx)$$

They are called fictitious because these forces do not arise due to interactions between particles but are due to the nature of a non-inertial frame.

Consider,

(1C.10) $[\dot{r}] \cdot F_{Cor} = (i\dot{x} + j\dot{y}) \cdot 2m\omega(i\dot{y} - j\dot{x})$

$$= 2m\omega(\dot{x}\dot{y} - \dot{y}\dot{x}) = 0$$

The Coriolis force is perpendicular to the velocity.

Also consider,

(1C.11) $r \cdot F_{trans} = (ix + jy) \cdot m\dot{\omega}(iy - jx) = 0$

The transverse force is perpendicular to the position vector.

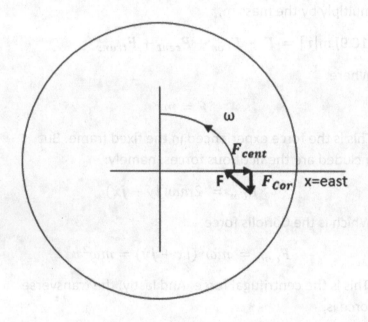

Fig. 1C.5

Example: Suppose a plane is flying east at a constant speed $[\dot{r}] = v_0$. The earth is rotating west to east. Looking down from the North Pole, we see a disk rotating counter-clockwise as shown in Fig. 1C.5.

We have ($\dot{\omega} = 0$),

(1C.12) $m[\ddot{r}] = 0 = F + F_{cent} + F_{Cor}$

$$= F + m\omega^2(ix) + 2m\omega(-jv_0)$$

There is no transverse force since ω is constant. We see that the centrifugal force acting towards the east (+i), while the Coriolis force is acting southward (−j). The force **F** represents the force of the earth exerting on the flying plane. Those three forces add up to zero. To move exactly to the east, as initially intended, the pilot has to fly slightly northward to offset the southward pull of the Coriolis force.

Example: Newton's law of gravity states that a force of attraction exists between any two masses, m_1 and m_2, at a distance r, and is given as,

(1C.13) $F_{gravity} = -\dfrac{Gm_1m_2}{r^2}$

Where G is a universal constant. In the MKS, it has the value of 6.674 x $10^{-11}m^3kg^{-1}s^{-2}$.

Consider the earth rotating around the sun in a circular path - the actual path is elliptical, but we are looking at a close approximation. The force of gravity plays the role of a centripetal force, keeping the earth in its orbit. We get,

$\rightarrow \dfrac{GM_{sun}M_{earth}}{R_{sun-earth}^2} = \dfrac{4\pi^2 R_{sun-earth}}{T^2}$ (equation 1A.16b)

We can rewrite this as,

$\rightarrow \dfrac{R^3}{T^2} = K$, where $K = \dfrac{GM_{sun}M_{earth}}{4\pi^2}$

Which is Kepler's 3rd law for the planets orbiting the sun.

Finally we have Newton's third law of motion.

The third law states that for every action there is an equal and opposite reaction.

But first we define the momentum as,

(1C.14) $p = mv$

More generally, the second law can also be written as,

(1C.15) $F = m \frac{dv}{dt} = \frac{d(mv)}{dt} = \frac{dp}{dt}$

If we consider a small time interval, Δt, we write,

(1C.16) $F\Delta t = \Delta p$

In the case of two billiard balls colliding, ball 1 will exert a force F_{12}. That is,

(1C.17) $F_{12}\Delta t_{12} = \Delta p_2$

Also, ball 2 exerts a force F_{21} on ball 1,

(1C.18) $F_{21}\Delta t_{21} = \Delta p_1$

According to Newton's 3rd law of motion, these two forces are equal and opposite,

(1C.19) $F_{12} = -F_{21}$

And since the time interval in which the two balls exert force on each other are equal ($\Delta t_{12} = \Delta t_{21}$). Then we get,

(1C.20) $\Delta p_1 = -\Delta p_2$

Expressing the change fully, we have

(1C.21) $m_1 v_1' - m_1 v_1 = -(m_2 v_2' - m_2 v_2)$

Where the prime indicates velocities after collision, and the unprimed indicates before collision. Rearranging, we get,

(1C.22) $m_2v_2 + m_1 v_1 = m_1v_1' + m_2v_2'$

This expresses the conservation law of momentum: the total momentum before collision = the total momentum after collision.

1D. Angular Momentum and Torques

Under a central force, we have:

(1D.1) $F = f(r)\dfrac{r}{r} = f(r)\lambda_r$

Where λ_r is a unit radial vector (Fig. 1B.1).

We start with the general equation of motion (eom) for a particle (equation 1C.1).

(1D.2) $F = ma$

Take the cross product (appendix D, equation DA.4a) of both sides with r:

(1D.3) $r \: X \: F = r \: X \: ma$

Consider:

(1D.4) $\dfrac{d}{dt}(r \: X \: mv) = v \: X \: mv + r \: X \: ma = r \: X \: ma$

Where the first term $v \: X \: mv = 0$, (see appendix D, equation DA.4a).

And this is the RHS of equation 1D.3.

(1D.3) $r \: X \: F = \dfrac{d}{dt}(r \: X \: mv)$

The cross product $r \: X \: mv$ is the angular momentum of the particle about the origin (Fig. 1D.1) denoted by J. The above equation 1D.3 now reads as:

(1D.4a) $r \, X \, F = \frac{d}{dt}(r \, X \, mv) = \dot{J}$

(1D.4b) Again $J = r \, X \, mv = r \, Xp$ is the angular momentum

The cross product $r \, X \, F$ defines the torque τ. We have,

(1D.4c) $\tau = \dot{J}$

In the absence of a torque, we get:

(1D.5) $\dot{J} = 0 \;\; \rightarrow \;\; J = constant$

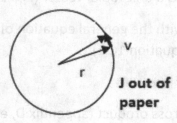

J out of paper

Fig. 1D.1

A consequence of this result is that in a central field, because J is perpendicular to both r and v then J is normal to the plane of motion (Fig. 1D.1).

Consider the the θ-direction in Fig. 1B.2. we have for the velocity in that direction

(1D.6) $v_\theta = r\dot{\theta}$ (equation 1B.3) = rω (equation 1C.3)

The angular momentum on this case (F=0),

(1D.7)$|J| = |r \, X \, mv|$

$$= rmv \text{ (since r is perpendicular to v).}$$

$$= mr^2\omega \text{ (equation 1D.6)}$$

42

Example: levers

Consider a balanced lever around a fulcrum (Fig 1D.2).

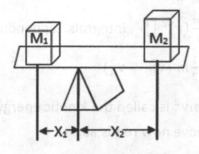

Fig. 1D.2

In both cases, the force (weight = mg) is perpendicular to the distance to the fulcrum, the torque is simply $(M_1g)X_1$ on the left, and $(M_2g)X_2$ on the right. Since the lever is balanced then the torques on each side are equal to each other:

(1D.8) $(M_1g)X_1 = (M_2g)X_2 \rightarrow \dfrac{M_1}{M_2} = \dfrac{X_2}{X_1}$

1E. Energy

Another important conservation law is that of energy. Before doing that, we define work as the product of a force applied on a body times the distance the force is applied, say from point A to point B:

(1E.1) $W = \int_A^B F \cdot dr$

$\quad = m \int_A^B \ddot{r} \cdot \dot{r} dt$, equations 1A.2, 1C.1

$\quad = m \int_A^B \dfrac{d\dot{r}}{dt} \cdot \dot{r} dt$, equation 1A.4

$$= \frac{m}{2} \int_A^B \frac{d}{dt} (\dot{r} \cdot \dot{r}) dt, \text{ Leibnitz rule (appendix E)}$$

$$= \frac{m}{2} \int_A^B d(\dot{r} \cdot \dot{r}),$$

$$= \frac{m}{2} (\dot{r})^2 \Big|_A^B, \text{ integrals, (appendix E)}$$

$$= \frac{1}{2} m (v_B^2 - v_A^2)$$

The quantity $\frac{1}{2} mv^2$ is called the kinetic energy (T) of a particle. The above now reads as,

(1E.2) $W = \frac{1}{2} m (v_B^2 - v_A^2) = T(B) - T(A)$

The work done on a body is equal to the change in the kinetic energy of the body as it moves from point A to point B.

If F is a conservative force, meaning over a closed path, the work is zero,

(1E.3) $W = \oint F \cdot dr = 0$

Then we can write the force as the negative gradient of a potential,

(1E.4) $F = -\nabla V$

We redo the steps in equation 1E.1 with 1E.4,

(1E.5) $W = \int_A^B F \cdot dr$

$$= - \int_A^B \nabla V \cdot dr \quad (\nabla V = \frac{dV}{dr})$$

$$= -\{V(B) - V(A)\} = V(A) - V(B)$$

Equating equations 1E.2 and 1E.5,

(1E.6) $V(A) - V(B) = T(B) - T(A)$

Or $\quad T(A) + V(A) = T(B) + V(B)$

This now reads as the followings: as the particle moves from point A to point B, through a conservative force, the sum of its kinetic energy and its potential energy remains constant. Hence the conservation law of total energy.

1F. Wave Motion

We begin with Hooke's law which states that the force needed to expand or compress a spring a certain distance is linearly proportional to that distance (Fig. 1F.1).

According to Newton's 2nd law of motion, we write,

(1F.1) $m\ddot{x} = -kx$

Where k is a constant characteristic of the spring.

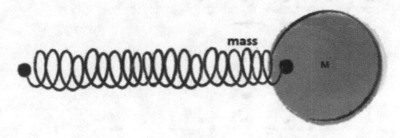

mass

M

Fig. 1F.1

We rewrite this as,

(1F.2) $\ddot{x} = -\omega^2 x$

Where $\omega = \sqrt{\dfrac{k}{m}}$. A solution of that equation is,

(1F.3) $x(t) = A \cos(\omega t + \varphi)$

Where A is the amplitude, and φ is a phase factor. Moreover, ω is the angular frequency, related to the frequency v (cycles per second) as

(1F.4) $\omega = 2\pi v$

The inverse of v is the period T, given in seconds per cycle.

(1F.5) $T = \dfrac{1}{v}$

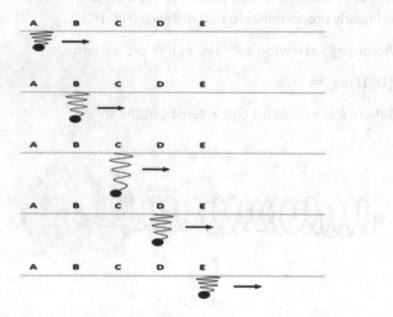

Fig. 1F.2

Oscillatory motion is narrowly linked to wave motion. Just consider a spring oscillating vertically, moving in the horizontal direction (Fig. 1F.2) through positions A to E.

Fig. 1F.3 depicts some of those points, which constitute part of the solution in equation 1F.3.

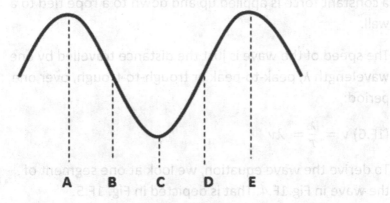

A   B   C   D   E

Fig. 1F.3

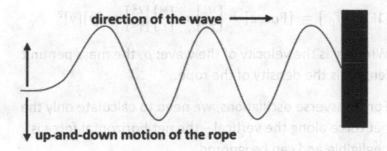

direction of the wave ⟶

up-and-down motion of the rope

Fig. 1F.4

Suppose the mass at the end of the spring is equipped with a pencil that can leave a mark on a piece of paper as it is moving from left to right.

The vibrations in the spring make up what is called longitudinal waves: the amplitude is in the same direction as the motion. Transversal waves occur when the amplitude is orthogonal to the motion (Fig. 1F.4) in which a constant force is applied up and down to a rope tied to a wall.

The speed of the wave is just the distance travelled by one wavelength λ, peak-to-peak or trough-to-trough, over one period:

(1F.6) $v = \frac{\lambda}{T} = \lambda\nu$

To derive the wave equation, we look at one segment of the wave in Fig.1F.4. That is depicted in Fig. 1F.5.

The tension T_0 is the force applied throughout the rope. We define the density ρ as the mass m per unit length. An analysis of the unit dimensions reveals the following:

(1F.7) $[T_0] = [\text{Force}] = \left[\frac{ML}{T^2}\right] = \left[\frac{M}{L}\right]\left[\frac{L^2}{T^2}\right] = [\rho][v]^2$

Where v is the velocity of the wave; ρ, the mass per unit length, is the density of the rope.

For transverse oscillations, we need to calculate only the net force along the vertical – the net horizontal force is negligible and can be ignored.

The vertical force at the point (x+dx, y+dy) is $T_0 \frac{\partial y}{\partial x}$ acting upward, evaluated at that point, while the vertical force at (x,y) is $T_0 \frac{\partial y}{\partial x}$ acting downward, evaluated at that point. The net force is then (appendix E),

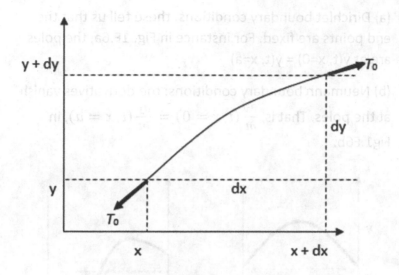

Fig. 1F.5

(1F.8) $dF_{vertical} = (T_0 \frac{\partial y}{\partial x})_{x+dx} - \left(T_0 \frac{\partial y}{\partial x}\right)_x = T_0 \frac{\partial^2 y}{\partial x^2} dx$

Newton's 2nd law states that the net force equals mass times acceleration,

(1F.9) $T_0 \frac{\partial^2 y}{\partial x^2} dx = m \frac{\partial^2 y}{\partial t^2} = \rho dx \frac{\partial^2 y}{\partial t^2}$

Cancel dx on both sides. Using 1F.7 for T_0, ρ cancels out, and we get

(1F.10) $\frac{\partial^2 y}{\partial x^2} = \frac{1}{v^2} \frac{\partial^2 y}{\partial t^2}$

This is known as the wave equation.

To find solutions to the wave equations, we must consider two types of boundary conditions:

(a) Dirichlet boundary conditions: these tell us that the end points are fixed. For instance in Fig. 1F.6a, the poles are at y(t, x=0) = y(t, x=a)

(b) Neumann boundary conditions: the derivatives vanish at the poles. That is, $\frac{\partial y}{\partial t}(t, x = 0) = \frac{\partial y}{\partial t}(t, x = a)$, in Fig1.F6b.

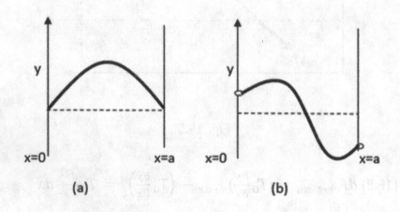

(a)   (b)

Fig. 1F.6

Consider the solution,

(1F.11) $y(x, t) = A(x) \cos(\omega t + \varphi)$

Take equation 1F.11, and twice the derivative with respect to x:

(1F.12) $\frac{\partial^2 y}{\partial x^2} = \cos(\omega t + \varphi)\frac{\partial^2 A(x)}{\partial x^2}$

Again take the same equation, and now twice the derivative with respect to t:

(1F.13) $\frac{\partial^2 y}{\partial t^2} = -A(x)\,\omega^2\cos(\omega t + \varphi)$

50

To satisfy the wave equation (1F.10), we need

(1F.14) $\dfrac{\partial^2 A(x)}{\partial x^2} = -\dfrac{\omega^2}{v^2} A(x)$

This is also a wave equation but in space rather than in time. We define the ratio,

(1F.15) $k \equiv \dfrac{\omega}{v} = \dfrac{2\pi v}{\lambda v} = \dfrac{2\pi}{\lambda}$

Where we used equations 1F.4 and 1F.6. And k is called the wave number.

We can also envision the most general solution to the wave equation to be:

(1F.16) $y(x,t) = \cos(\omega t + \varphi)\{A\sin\left(\dfrac{2\pi x}{\lambda}\right) + B\cos\left(\dfrac{2\pi x}{\lambda}\right)\}$

Consider that the string is fixed at both ends for all times.

At x =0, we have

(1F.17) $y(0,t) = \cos(\omega t + \varphi)\{0 + B\} = 0$

$\rightarrow B = 0$

Equation 1F.16 becomes

(1F.18) $y(x,t) = A\cos(\omega t + \varphi)\sin\left(\dfrac{2\pi x}{\lambda}\right)$

To satisfy the fixed point at x = a, y(a,t) = 0, we don't want to choose the trivial solution A = 0. The only way out is to choose

(1F.19) $\sin\left(\dfrac{2\pi a}{\lambda}\right) = 0$

This contains multiple solutions:

(1F.20) $\frac{2\pi a}{\lambda} = 0, \pi, 2\pi, 3\pi$ ...

Note: the first case, $\frac{2\pi a}{\lambda} = 0$, means that $\lambda = \infty$, that is, we have a straight line segment.

We labels these normal modes starting with the number 1, and 2, and so on.

(1F.21) $\lambda_1 = 2a$, $\lambda_2 = \lambda_1$, $\lambda_3 = \frac{1}{3}\lambda_1$ ...

Using complex numbers, the most general solution of the wave equation is,

(1F.22) $y(x, t) = A\, e^{i(kx-wt)}$

Where A is an arbitrary number to be determined by boundary conditions.

Example: potential energy of a spring

The work done by a spring is, (equation 1E.1)

(1F.23) $W = \int_A^B F \cdot dr$

Substitute the force (equation 1F.1) in stretching the spring from 0 to x,

$\rightarrow W = \int_0^x kx\, dx$

Note the force is taken to be positive because in stretching the spring, we are doing work against the spring, storing potential energy into the spring. Integrating (Equation EB.5), we can write,

(1F.24) $V_{spring} = \frac{1}{2}kx^2$

Where V_{spring} is the potential energy stored in the spring.

1G. Fields

Consider the case of a source and a test particle, the understanding is that the particle's mass is much smaller than the mass of the source ($M_{source} \gg m_{test}$). We define the gravitational field as,

(1G.1) $g \equiv \dfrac{F_{gravity}}{m_{test}}$

Recall Newton's law of gravity which states that a force of attraction exist between any two masses, m_1 and m_2, at a distance r and is given as,

(1G.2) $F_{gravity} = -\dfrac{Gm_1m_2}{r^2}$ (equation 1C.13)

Substitute into the above for a test particle just above the surface of the earth (Fig. 1G.2), we get for the magnitude of the field on the surface of the earth,

(1G.3) $g = \dfrac{GM_{earth}\,m_{test}}{R_{earth}^2}\,\dfrac{1}{m_{test}} = \dfrac{GM_{earth}}{R_{earth}^2}$

Taking $M_{earth} = 5.98 \times 10^{24} kg$, $R_{earth} = 6.36 \times 10^6 m$

$\rightarrow g = \dfrac{6.674 \times 10^{-11} m^3 kg^{-1}s^{-2} \times 5.98 \times 10^{24} kg}{(6.36 \times 10^6 m)^2}$

$\rightarrow \quad = \dfrac{39.9 \times 10^{13}}{40.4 \times 10^{12}}\dfrac{m}{s^2} = 9.87\ m/s^2$

This is what Galileo had observed: that the acceleration of a test particle due to the gravity of the earth is independent of the mass of the test particle. Its value has been overwhelmingly confirmed. This was a triumph of Newton's theory of gravity.

(1G.4) Gauss' Law of Gravity

Note: for the gravitational field, we use g for the field near the surface of the earth, and ϕ for the general case.

Consider the case in which the sum of all the particles can be replaced by a single mass M (where V ≡ volume)

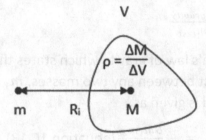

Fig. 1G.1

The total intensity of the gravitational field ϕ over a closed surface V, according to Newton's law (1G.3) is:

$$\int_V \phi \cdot dS = - \int_V \frac{GM}{R^2} e_r \cdot dS$$

Where e_r is a radial unit vector.

Now the magnitude of the infinitesimal area element dS is just the area of the infinitesimal solid angle $d\Omega$, given by,

$$dS = R^2 d\Omega \, e_r$$

Substituting,

$$\int_V \boldsymbol{\phi} \cdot dS = - \int_V \frac{GM}{R^2} e_r \cdot (R^2\, e_r\, d\Omega)$$

$$= - GM \int_V e_r \cdot e_r\, d\Omega$$

$$= - GM \int_V d\Omega, \quad (e_r^2 = 1)$$

$$= - GM\, 4\pi$$

Now we can substitute for the mass M,

$$M = \int_V \rho\, dV$$

$$\int_V \boldsymbol{\phi} \cdot dS = - 4\pi G \int_V \rho\, dV$$

For the left-hand side, using Gauss theorem (appendix E, equation EC.1) we get,

$$\int_V dV\, \nabla \cdot \boldsymbol{\phi} = - 4\pi G \int_V \rho\, dV$$

Since both sides are integrations over an arbitrary volume V, we can conclude that,

$$\nabla \cdot \boldsymbol{\phi} = - 4\pi G \rho$$

This is known as Gauss' Law, which is another version of Newton's Law of Gravity.

(1G.5) Example: The potential energy in a gravitational field.

Excluding the kinetic energy, the gravitational field applies a force of

(1) $F = mg = -\frac{Gm_1m_2}{r^2} e_r$

The work done by this force is then (equation 1E.1),

(2) $dW = F \cdot dr = -\frac{Gm_1m_2}{r^2} e_r \cdot dr = -\frac{Gm_1m_2}{r^2} dr$

Integrate to get the total work done,

(3) $W = -Gm_1m_2 \int_{r_1}^{r_2} \frac{dr}{r^2} = -Gm_1m_2(\frac{1}{r_2} - \frac{1}{r_1})$

Consider the work done on a test particle starting at infinity to a distance r from the source of gravity, that is, we take the ground zero of potential energy to be at infinity (Fig. 1G.2).

In the case of the earth acting as the source on a test particle at a distance r, we then have,

(4) $\frac{1}{r_1 = \infty} = 0, \quad \frac{1}{r_2} = \frac{1}{r}$

(5) $W = \Delta V = -\frac{GM_{earth}m_{test}}{r}$

For an object above the ground at a height h (Fig. 1G.2), $r = R_{earth} + h$, in which case $R_{earth} \gg h$, we have

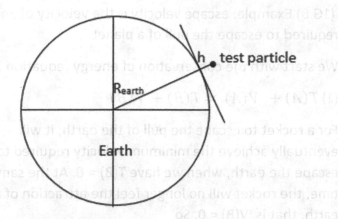

Fig. 1G.2

(6) $W = -\dfrac{GM_{earth}m_{test}}{R_{earth} + h} = -\dfrac{GM_{earth}m_{test}}{R_{earth}(1 + \frac{h}{R_{earth}})}$

$= -\dfrac{GM_{earth}m_{test}}{R_{earth}}(1 + \dfrac{h}{R_{earth}})^{-1}$

$\approx -\dfrac{GM_{earth}m_{test}}{R_{earth}}(1 - \dfrac{h}{R_{earth}})$

$= -\dfrac{GM_{earth}m_{test}}{R_{earth}} + \dfrac{GM_{earth}\,m_{test}}{R_{earth}^2}h$

Consider the second term:

$\dfrac{GM_{earth}\,m_{test}}{R_{earth}^2}h = (\dfrac{GM_{earth}}{R_{earth}^2})m_{test}h = mgh$ (equation 1G.3)

Where we drop the subscript "test". Now what is important is differences in potential energy.

(7) $\Delta V = V_f - V_i$

$= -\dfrac{GM_{earth}m_{test}}{R_{earth}} + mgh - (-\dfrac{GM_{earth}m_{test}}{R_{earth}})$

$= mgh$

(1G.6) Example: escape velocity is the velocity of an object required to escape the pull of a planet.

We start with the conservation of energy (equation 1E.6)

(1) $T(A) + V(A) = T(B) + V(B)$

For a rocket to escape the pull of the earth, it will eventually achieve the minimum velocity required to escape the earth, when we have T(B) = 0. At the same time, the rocket will no longer feel the attraction of the earth, that is, V(B) = 0. So

(2) $\frac{1}{2}mv^2 - \frac{GMm}{r} = 0$ (equations 1E.2 and 1G.5)

Where m is the mass of the rocket, M is the mass of the earth, and r is the distance between the rocket and the center of the earth (radius of the earth).

Divide by m, and simplify.

(3) $v^2 = \frac{2GM}{r}$

Or

(4) $v = \sqrt{\frac{2GM}{r}}$

Note that this result is independent of the mass (m) of the rocket.

Again taking $M_{earth} = 5.98 \times 10^{24}kg$, and $R_{earth} = 6.36 \times 10^6 m$, we get

(5) $v = \sqrt{\frac{2(6.674 \times 10{-11}m^3kg^{-1}s^{-2})5.98 \times 10^{24}kg}{6.36 \times 10^6 m}}$

$\approx 11200 ms^{-1}$

= 11.2 Km/s or 40,320 Km/hr.

Chapter 2

Principle of Least Action

2A. Lagrangian formalism

An object S is said to be invariant if a change in S vanishes. That is, mathematically, if S is invariant then $\delta S = 0$.

The Lagrangian is defined as the difference between the kinetic energy and the potential energy of a particle.

(2A.1) $L = T - V$

Note that the Lagrangian is not necessarily conserved. What is conserved is the total energy, called the Hamiltonian,

(2A.2) H = T + V.

Generally, the Lagrangian is a function of position and velocity, while the Hamiltonian is a function of position and momentum.

(2A.3) $L \rightarrow L(q, \dot{q}) \qquad H \rightarrow H(q, p)$

Where the dot indicates a derivative with respect to time.

The Action is defined in terms of the Lagrangian as:

(2A.4) $S = \int L(q, \dot{q}) \, dt$

The principle of least action states that a particle will follow the path of least action. To find this path we minimize the variance of the action:

(2A.5) $\delta S = 0$

Substituting equation 2A.4 into the above:

(2A.6) $\int \delta L(q, \dot{q}) dt = 0$

Taking into consideration that we may have many particles, the LHS becomes,

(2A.7) $\int \delta L(q, \dot{q}) dt = \int \sum_i \left[\frac{\partial L}{\partial q_i} \delta q_i + \frac{\partial L}{\partial \dot{q}_i} \delta \dot{q}_i \right] dt$

(See appendix E on partial derivatives and integration by parts.)

Integrating by parts the second term in the bracket, we get

(2A.8) $\delta S = \int \sum_i \left[\frac{\partial L}{\partial q_i} - \frac{d}{dt} \left(\frac{\partial L}{\partial \dot{q}_i} \right) \right] \delta q_i \, dt$

Since q_i is arbitrary, the above expression vanishes only if the term inside the bracket vanishes,

(2A.9) $\frac{\partial L}{\partial q_i} - \frac{d}{dt} \left(\frac{\partial L}{\partial \dot{q}_i} \right) = 0$

These are called the Euler-Lagrange equations. What's so important about those equation is that they yield the equations of motion (eom).

EXAMPLE: take a single particle moving along the x direction with kinetic energy L = $\frac{1}{2} m\dot{x}^2$, into a potential V(x).

From equation 2A.1, the Lagrangian is,

(2A.10) $L = \frac{1}{2} m\dot{x}^2 - V(x)$

From 2A.9, we want to calculate the first term. We take the derivative of L with respect to x:

(2A.11) $\frac{\partial L}{\partial x} = \frac{\partial}{\partial x} \left[\frac{1}{2} m\dot{x}^2 - V(x) \right] = -\frac{\partial}{\partial x} V(x)$

To calculate the second term, we first take the derivative of L with respect to $\dot{x}$,

(2A.12) $\frac{\partial L}{\partial \dot{x}} = \frac{\partial}{\partial \dot{x}}[\frac{1}{2}m\dot{x}^2 - V(x)] = m\dot{x}$

Next we take the time derivative,

(2A.13) $\frac{d}{dt}\left(\frac{\partial L}{\partial \dot{x}}\right) = m\ddot{x}$

Putting everything together, we get

(2A.14) $\frac{\partial}{\partial x}V(x) + m\ddot{x} = 0$

A conservative force can be expressed in terms of a potential (equation 1E.4):

(2A.15) $F = -\frac{\partial}{\partial x}V(x)$

Now $\ddot{x}$ *is the accelation*, so then equation 2A.14 is just Newton's 2nd law of motion, F = ma.

The conjugate momentum is defined as,

(2A.16) $p = \frac{\partial L}{\partial \dot{x}}$

In the above example, from equation 2A.12, p = m$\dot{x}$ = mv, which is the classical momentum.

Another important role of the Lagrangian is its relationship to conservation law. Consider a particle moving in a plane, rotating about the axis perpendicular to this page:

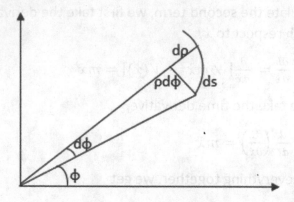

Fig. 2A.1

In this case the Lagrangian is just the kinetic energy, (V(x) =0 in 2A.1, that is, no force is acting on the particle).

(2A.17) $L = \frac{1}{2} m v^2$

However in this case, it's better to work in polar coordinates. The distance travelled (ds) by the particle along the arc s in time (dt) is (Pythagoras theorem, equation CB.3),

(2A.18) $ds = [(d\rho)^2 + (\rho d\varphi)^2]^{\frac{1}{2}}$

Note: lower case s for the arc interval, upper case S for the action.

The magnitude of the speed is then,

(2A.19) $v = \frac{ds}{dt} = [(\dot\rho)^2 + (\rho\dot\varphi)^2]^{\frac{1}{2}}$

From this we get the Lagrangian,

(2A.20) $L = \frac{1}{2}m\left[(\dot{\rho})^2 + (\rho\dot{\varphi})^2\right]$

We now calculate the equation of motion (from 2A.9).

For $q_1 = \rho$

(2A.21) $\frac{\partial L}{\partial \rho} - \frac{d}{dt}\left(\frac{\partial L}{\partial \dot{\rho}}\right) = 0 \quad \rightarrow \quad m\rho\dot{\varphi}^2 - \frac{d}{dt}(m\dot{\rho}) = 0$

For $q_2 = \varphi$

(2A.22) $\frac{\partial L}{\partial \varphi} - \frac{d}{dt}\left(\frac{\partial L}{\partial \dot{\varphi}}\right) = 0 \quad \rightarrow \quad 0 - \frac{d}{dt}(m\rho^2\dot{\varphi}) = 0$

We can read this better if we take the momentum (equation 2A.16),

(2A.23) $\quad p_\rho = \frac{\partial L}{\partial \dot{\rho}} = m\dot{\rho} \qquad p_\varphi = \frac{\partial L}{\partial \dot{\varphi}} = m\rho^2\dot{\varphi}$

The above equations become,

(2A.24) $\frac{dp_\rho}{dt} = m\rho\dot{\varphi}^2\,; \qquad \frac{dp_\varphi}{dt} = 0$

The first equation gives the centripetal force, a fictitious force acting along the radial axis (equation 1C.9). The second says that the angular momentum is conserved when no force is applied on the particle (equation 1D.7).

2B. Hamiltonian Formalism

The Hamiltonian is defined through the Legendre transform (Appendix E, equation EA.22),

(2B.1) $H = p\dot{x} - L$

In the above example, equation 2A.10, we have

(2B.2) $H = (m\dot{x})\dot{x} - (\frac{1}{2}m\dot{x}^2 - V(x))$

$$= \tfrac{1}{2}m\dot{x}^2 + V(x) = \text{T+V}$$

And that is the total energy of the system.

Checking the dimension units of certain physical objects (appendix B):

a) The Hamiltonian: $[H] = [L] = [M] = 1$

b) The action: $[S] = [MT] = 0$

In Hamiltonian mechanics, a classical physical system is described by a set of canonical coordinates $r = (q, p)$, where each component of the coordinate q_i (position) and p_i (momentum) is indexed to the frame of reference of the system. The time evolution of the system is uniquely defined by Hamilton's equations.

The equation of motion are given as,

(2B.3) $\dfrac{dp}{dt} = -\dfrac{\partial H}{\partial q}$; $\dfrac{dq}{dt} = \dfrac{\partial H}{\partial p}$

From equation 2B.1 ($q \equiv x$), this gives

(2B.4a) $\partial H/\partial p = \dot{x}$

(2B.4b) $\partial H/\partial q = -\partial L/\partial q$ (equation 2B.1)

$$= -\frac{\partial}{\partial x}(T - V) \text{ (equation 2A.1)}$$

$$= \frac{\partial}{\partial x}V(x) = -F \quad \text{(equation 2A.15)}$$

$$= -\frac{dp}{dt} \qquad \text{(equation 1C.15)}$$

The advantage is that the Hamiltonian equations are first order, while the Euler-Lagrange equations are second order.

Define the Poisson bracket (using the { } bracket notation),

(2B.5) $\{A,B\} = (\partial A/\partial x)(\partial B/\partial p) - (\partial A/\partial p)\partial B/\partial x$,

where A and B are any two functions.

In particular, if B = H,

(2B.6) $\{A,H\} = (\partial A/\partial x)(\partial H/\partial p) - (\partial A/\partial p)(\partial H/\partial x)$

$\quad = (\partial A/\partial x)(dx/dt) + (\partial A/\partial p)(dp/\partial t)$

Where we used equation 2B.3.

Now, $A \rightarrow A(x,p)$

$dA = (\partial A/\partial x)(dx) + (\partial A/\partial p)(dp)$ (appendix E, equ. EA.18)

Or

$dA/dt = (\partial A/\partial x)(dx/dt) + (\partial A/\partial p)(dp/dt)$

Then $dA/dt = \{A,H\}$ (equation 2B.6)

The equations of motion are further simplified as,

(2B.7a) $dx/dt = \{x, H\} = \partial H/\partial p$ (equ. 2B.3)

(2B.7b) $dp/dt = \{p, H\} = -\partial H/\partial p$ (equ. 2B.3)

For the position (A = x) and the momentum (B = p), the Poisson bracket becomes,

(2B.8) $\{x,p\} = 1$

As we will see later on, to cross the threshold from

Classical Physics to Quantum Mechanics, we replace that Poisson bracket with the commutation relation (see chapter 7, equation 7A.13a for the definition of a commutation),

(2B.9) $[x,p] = i\hbar$,

which yields the Heisenberg Uncertainty Principle!!!

2C. Classical Field Theory

We want to develop similar Lagrangian and Hamiltonian for fields. We take our cues from the Maxwell equations, which we will see in greater details in chapter 3. But for now let us say that the electric field E_r and magnetic field B_r (where r = 1,2,3 for x,y,z) is replaced by the electromagnetic field A_μ, (where μ = 0,1,2,3, with μ = 0 ≡ t, the time component) such that

(2C.1) $E_r = \partial A_r/\partial t - \partial A_0/\partial x^r$

(2C.2) $B_r = \frac{1}{2} \epsilon_{rst}\partial A_t/\partial x^s$

It's not important to know how we obtain these fields but to notice that for the general case of any field $\varphi(x,t)$, the Lagrangian will now be a function of the field itself $\varphi(x,t)$, and its derivatives - the time derivative $\partial\varphi(x,t)/\partial t$ and the spatial derivative $\nabla\varphi(x,t)$. Therefore we define,

(2C.3) $L(t) = \int dx^3 \mathcal{L}(\varphi_a,\partial_\mu\varphi_a)$

Where $\mathcal{L}(\varphi_a,\partial_\mu\varphi_a)$ is now the Lagrangian density.

The Action is then

(2C.4) $S = \int dt L(t) = \int dx^4\, \mathcal{L}(\varphi_a, \partial_\mu \varphi_a)$

Where $dx^4 = dtdxdydz$, is the 4-D volume.

We still determine the equations of motion for fields by the principle of least action, that is,

(2C.5) $\delta S = 0$, keeping the end points fixed.

In the same vein as equation 2A.7, we have

(2C.6) $\delta S = \int dx^4\, \{(\partial\mathcal{L}/\partial\varphi_a)\delta\varphi_a + (\partial\mathcal{L}/\partial(\partial_\mu\varphi_a))\delta(\partial_\mu\varphi_a)\}$

$= \int dx^4[\{\partial\mathcal{L}/\partial\varphi_a - \partial_\mu(\partial\mathcal{L}/\partial(\partial_\mu\varphi_a))\}\delta\varphi_a$

Where the last term was obtained by integration by parts (appendix E, equation EB.4). For $\delta S = 0$ the term inside the bracket must vanish.

(2C.7) $\partial_\mu(\partial\mathcal{L}/\partial(\partial_\mu\varphi_a)) - \partial\mathcal{L}/\partial\varphi_a = 0$

These are called the Euler-Lagrange equations for fields.

2D. Noether Theorem

Consider a set of scalar fields φ_a and a Lagrangian $\mathcal{L}(\varphi_a, \partial_\mu\varphi_a)$. Suppose we make an infinitesimal change,

(2D.1) $\varphi_a(x) \rightarrow \varphi_a(x) + \delta\varphi_a(x)$

(2D.2) $\mathcal{L}(x) \rightarrow \mathcal{L}(x) + \delta\mathcal{L}(x)$

Expanding $\delta\mathcal{L}(x)$ in terms of its arguments,

(2D.3) $\delta\mathcal{L}(x) = \frac{\partial\mathcal{L}}{\partial\varphi_a(x)}\delta\varphi_a(x) + \frac{\partial\mathcal{L}}{\partial(\partial_\mu\varphi_a(x))}\partial_\mu\varphi_a(x)$

Reiterating: the classical equations of motion are given by the principle of least action:

(2D.4) $\frac{\delta S}{\delta\varphi_a(x)} = 0$

Where

(2D.5) $S = \int d^4y\, \mathcal{L}(y)$

Take the derivative of 2D.5 with respect to $\varphi_a(x)$,

(2D.6) $\frac{\delta S}{\delta\varphi_a(x)} = \int d^4y\, \frac{\mathcal{L}(y)}{\delta\varphi_a(x)}$

Expanding $\mathcal{L}(y)$ (equation 2D.2),

(2D.7)

$$\frac{\delta S}{\delta\varphi_a(x)} = \int d^4y\, [\frac{\mathcal{L}(y)}{\delta\varphi_b(y)}\frac{\delta\varphi_b(y)}{\delta\varphi_a(x)} + \frac{\partial\mathcal{L}(y)}{\partial(\partial_\mu\varphi_b(y))}\frac{\delta\partial_\mu\varphi_b(y)}{\delta\varphi_a(x)}]$$

$$= \int d^4y\, [\frac{\mathcal{L}(y)}{\delta\varphi_b(y)}\delta_{ba}\delta^4(y-x)$$

$$+ \frac{\partial\mathcal{L}(y)}{\partial(\partial_\mu\varphi_b(y))}\delta_{ba}\delta^4\partial_\mu(y-x)]$$

Integrate the last term by parts (equation EB.4),

$$= \int d^4y\, [\frac{\mathcal{L}(y)}{\delta\varphi_b(y)}\delta_{ba}\delta^4(y-x)$$

$$- \partial_\mu\frac{\partial\mathcal{L}(y)}{\partial(\partial_\mu\varphi_b(y))}\delta_{ba}\delta^4(y-x)]$$

Using the Delta function (appendix J, equation JB.1) the full integration then is,

(2D.8) $\dfrac{\delta S}{\delta \varphi_a(x)} = \dfrac{\mathcal{L}(x)}{\delta \varphi_a(x)} - \partial_\mu \dfrac{\partial \mathcal{L}(x)}{\partial(\partial_\mu \varphi_a(x))}$

Re-arranging,

(2D.9) $\dfrac{\mathcal{L}(x)}{\delta \varphi_a(x)} = \partial_\mu \dfrac{\partial \mathcal{L}(x)}{\partial(\partial_\mu \varphi_a(x))} + \dfrac{\delta S}{\delta \varphi_a(x)}$

Substitute that into equation 2D.3

(2D.10) $\delta \mathcal{L}(x) = \dfrac{\partial \mathcal{L}}{\partial \varphi_a(x)} \delta \varphi_a(x) + \dfrac{\partial \mathcal{L}}{\partial(\partial_\mu \varphi_a(x))} \partial_\mu \varphi_a(x)$

$= \left[\partial_\mu \dfrac{\partial \mathcal{L}(x)}{\partial(\partial_\mu \varphi_a(x))} + \dfrac{\delta S}{\delta \varphi_a(x)} \right] \delta \varphi_a(x) + \dfrac{\partial \mathcal{L}}{\partial(\partial_\mu \varphi_a(x))} \partial_\mu \varphi_a(x)$

$= \partial_\mu \dfrac{\partial \mathcal{L}(x)}{\partial(\partial_\mu \varphi_a(x))} \delta \varphi_a(x) + \dfrac{\delta S}{\delta \varphi_a(x)} \delta \varphi_a(x)$

$\qquad\qquad + \dfrac{\partial \mathcal{L}}{\partial(\partial_\mu \varphi_a(x))} \partial_\mu \varphi_a(x)$

Combining first and last terms, using the Leibnitz rule (appendix E, equationEA.9b)

(2D.11) $\delta \mathcal{L}(x) = \partial_\mu \left(\dfrac{\partial \mathcal{L}(x)}{\partial(\partial_\mu \varphi_a(x))} \delta \varphi_a(x) \right) + \dfrac{\delta S}{\delta \varphi_a(x)} \delta \varphi_a(x)$

We define the Noether current as,

(2D.12) $j^\mu = \dfrac{\partial \mathcal{L}(x)}{\partial(\partial_\mu \varphi_a(x))} \delta \varphi_a(x)$

Equation 2D.11 can now be re-written as,

(2D.13) $\partial_\mu j^\mu = \delta \mathcal{L}(x) - \dfrac{\delta S}{\delta \varphi_a(x)} \delta \varphi_a(x)$

If the classical fields satisfy the equations of motion (2D.4) then $\delta S = 0$ and we get,

(2D.14) $\partial_\mu j^\mu = \delta \mathcal{L}(x)$

If the Lagrangian is invariant under the infinitesimal change ($\delta\mathcal{L}(x) = 0$) the Noether current is conserved.

(2D.15)) $\partial_\mu j^\mu = 0$

This is known as Noether theorem. Using the definition in appendix E of ∂_μ (equation EA.16),

(2D.16) $\partial_\mu \equiv \partial_t + \nabla$

We have,

(2D.17) $\partial_t j^0(x) + \nabla \cdot \mathbf{j}(x) = 0$

Where $j^0(x)$ is the charge density, and $\mathbf{j}(x)$ is the current density. Equation 2D.17 expresses the conservation of a local charge. This is found by integrating the time component of j^μ. First we define the Noether charge:

(2D.18) $Q = \int j^0 dV$

Where dV is a volume element. For the charge to be conserved its time derivative must be equal to zero.

(2D.18) $\frac{dQ}{dt} = \int \partial_t j^0 dV$

$= -\int \nabla \cdot \mathbf{j}\, dV$ (equation 2D.17)

$= -\int \mathbf{j} \cdot d\mathbf{S}$ (Gauss theorem, equation EC.1)

$= 0$

We can justify the last line by considering at infinity, the current $\mathbf{j}$ over the surface $\mathbf{S}$ dies off.

Chapter 3

Electromagnetism

3A. The Maxwell's equations

First we will illustrate the Maxwell's equations in terms of the electric and magnetic fields. Note: we are using generalized units for which c = 1, and all other constants are also set to 1.

<u>First Law,</u>

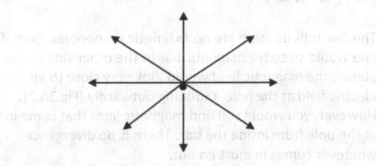

Fig. 3A.1

(3A.1) $\nabla \cdot E = \rho$

This law comes from Coulomb's law of electrostatics, which states that an electric force exists between charges (unlike charges attract, like charges repel), and similar to the force of gravity, an inverse squared law (equation 1C.13). Similarly, using Gauss's theorem (EC.1), we get equation 3A.1, where ρ is the electrical charge density. This law also tells us that charge densities produce electric fields, and those fields radiate outwardly (Fig. 3A.1).

Second Law

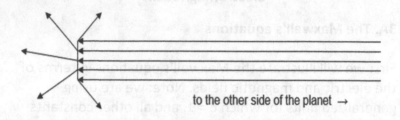

to the other side of the planet →

Fig. 3A.2

(3A.2) $\nabla \cdot B = 0$

This law tells us there are no magnetic monopoles. Even if one would stretch a magnetic bar to the other side of the planet, the magnetic field would look very close to an electric field at the pole, radiating outwardly (Fig.3A.2). However, you would still find magnetic lines that come in at the pole from inside the bar. There is no divergence: whatever comes in must go out.

Third law

(3A.3) $\nabla \times E = -\dot{B}$

This law tells us that a changing magnetic field will produce an electric field that wraps around this changing magnetic field (Fig.3A.3). The greater the change in the magnetic field being created, say by an increasing electric current, the stronger is the electric field. The negative sign indicates that the changing electric field being created will produce a counter magnetic field (from the fourth law) in opposition. This is a consequence of Newton's third law of motion: for every action, there is a counter-reaction.

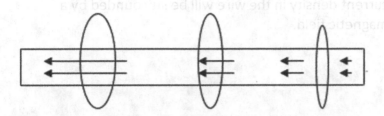

Fig. 3A.3

<u>Fourth law</u>

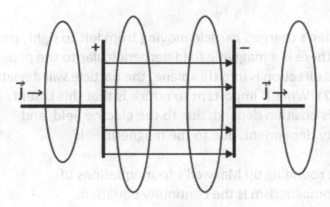

Fig. 3A.4

(3A.4) $\nabla \times B = \dot{E} + j$

This law tells us that a changing electric field or a current density will produce a magnetic field (Fig. 3A.4). For example, building charges on a capacitor will build an electric field between the plates, which will build a magnetic field that wraps around this changing electric field. Also, in building the charges on the plates, the current density in the wire will be surrounded by a magnetic field.

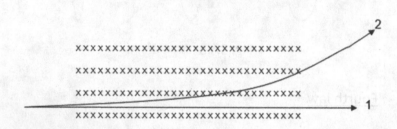

Fig. 3A.5

Consider a charged particle moving from left to right, path (1). If there is a magnetic field perpendicular to the paper, and its direction is into this plane, the particle will deviate, path (2). What is important to notice is that this Lorentz force is position depend, due to the electric field, and velocity dependent, due to the magnetic field.

Lastly, rounding up Maxwell's four equations of electromagnetism is the continuity equation.

The argument is as follows: we know that electric charge is conserved but what would prevent a charge from disappearing here on Earth and reappearing somewhere else, say on Mars? We would still have conservation but a description of the real world would have to take into account that if charges were moving out from a particular location, there would be a flow of these charges. And one could presumably place a surface around that location to detect that indeed the charges were flowing out. Another way of saying this is that if charges were moving out through the boundaries of a surface there would be an electric current passing through that surface. So now we want to formulize this concept in some mathematical frame.

We define the local charge density of region of space as the total charge per volume of that space. For a volume of small size ξ,

(3A.5) $\rho(x,t) = \dfrac{q}{\xi^3}$

Another way of putting this,

(3A.6) $q = \int \rho(x,t) \, d^3x$

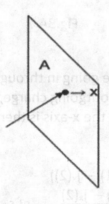

Fig. 3A.6

The current density passing through a certain region is the amount of charge going through per unit area per unit time. It is a vector, so we must take components. Consider the x direction to be perpendicular to the area (Fig. 3A.6).

Then the x-component of the current density is,

$$(3A.7)\ j_x\ =\ \frac{\delta q}{A\ \delta t}\ \ \rightarrow\ \ \delta q = A\ \delta t\ j_x$$

To calculate the total charge flowing through a volume of space, we must consider the charge going in and out at each of the surfaces of an enclosing box.

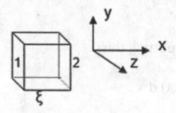

Fig.3A.7

The amount of charge going in through face 1 is $j_x(1)\ \xi^2\ \delta t$. At face 2, there is an outgoing charge, which is $j_x(2)\ \xi^2\ \delta t$. The net charge along the x-axis is then the difference (equation 3A.7),

$$(3A.8)\ \delta q\ =\ \xi^2\ \delta t\ (j_x(1) - j_x(2))$$

$$=\ \xi^3\ \delta t\ \frac{j_x(1) - j_x(2)}{\xi}\ =\ -\ \xi^3\ \delta t\ \frac{j_x(2) - j_x(1)}{\xi}$$

But the change of a quantity at two positions over a small interval is just the derivative of that quantity.

(3A.9) $\delta q = -\xi^3 \, \delta t \, \dfrac{\partial j_x}{\partial x}$

We get similar expressions along the y and z directions. We can summarize this as,

$\rightarrow \delta q = -\xi^3 \, \delta t \, (\dfrac{\partial j_x}{\partial x} + \dfrac{\partial j_y}{\partial y} + \dfrac{\partial j_z}{\partial z})$

Or,

(3A.10) $\dfrac{dq}{dt} = -\xi^3 \, (\dfrac{\partial j_x}{\partial x} + \dfrac{\partial j_y}{\partial y} + \dfrac{\partial j_z}{\partial z})$

Divide both sides by ξ^3 and then recall equation 3A.5 and the divergence (see equation DA.5).

(3A.11) $\dfrac{\partial \rho}{\partial t} = -\, \nabla \cdot \mathbf{j}$ (Compare to equation 2D.17)

This is called the *continuity equation*. We can rewrite it as,

(3A.12) $\dot{\rho} + \nabla \cdot \mathbf{j} = 0$

In the context of the 4-vector formalism (see chapter 5), we write $\rho = j^0$. Similarly, $j_x = j^1$, $j_y = j^2$, $j_z = j^3$. Equation 3A.12 takes on the simple form,

(3A.13) $\partial_\mu j^\mu = 0$ (see equation 2D.15)

A familiar exercise is to break up equations like 3A.13 into its temporal and spatial components:

(3A.14) $\partial_0 j^0 + \partial_i j^i = 0$

Where we use $t \rightarrow 0$

$x, y, z \rightarrow i = 1, 2, 3$

Note: Equations 3A.12, 3A.13 and 3A.14 are all equivalent.

3B. The Magnetic Field

We will now explore some features of the magnetic field. Consider a charged particle moving in a constant magnetic field. A suitable Lagrangian is,

(3B.1) $\mathcal{L} = \frac{1}{2} m \; \dot{x}_i^2 + q \; \dot{x}_i \; A_i(x_i)$

Recall that $v = \frac{dx}{dt} \equiv \dot{x}$

And the corresponding action is (equation 2A.4),

(3B.2) $S = \int_{t_1}^{t_2} dt \; (\frac{1}{2}mv^2) + q \int_{t_1}^{t_2} \mathbf{A} \cdot d\mathbf{x}$

Where **A**, known as the magnetic vector potential, is a function of the position, and q is the charge in some suitable units. The first term in 3B.1 is due to the kinetic energy; the second term, to the magnetic potential energy. It was obtained by the following:

$\rightarrow \int_{t_1}^{t_2} q \; \dot{x}_i \; A_i(x_i)dt = q \int_{t_1}^{t_2} \frac{dx_i}{dt} A_i dt = q \int_{t_1}^{t_2} \mathbf{A} \cdot d\mathbf{x}$

In order to convince ourselves that this is the right Lagrangian, we want to show by calculating the equations of motion what the force will turn out to be,

(3B.3) $\mathbf{F} = q\mathbf{v} \times \mathbf{B}$

Where

(3B.4) $\mathbf{B} = \nabla \times \mathbf{A}$

Note: the magnetic field **B** is associated with the vector potential **A**.

The Euler-Lagrange Equations are (from equations 2A.9 and 2A.16),

$$(3B.5) \quad \frac{dP_i}{dt} = \frac{\partial \mathcal{L}}{\partial x_i}$$

First, let's determine the canonical momentum (from equation 2A.16). The z-component is,

$$(3B.6) \quad P_z = \frac{\partial \mathcal{L}}{\partial \dot{z}} = \frac{\partial}{\partial \dot{z}} (\tfrac{1}{2} m \ \dot{z}_i^2 + q \ \dot{z} \ A_i(z_i) \ = m \dot{z} + q A_z$$

Taking the derivative with respect to time,

$$(3B.7) \quad \frac{dP_z}{dt} = m \ \ddot{z} + q \frac{dA_z}{dt}$$

This is the left-hand side of the Euler-Lagrange equation 3B.5. For the right hand-side, we must differentiate equation 3B.1 with respect to z,

$$(3B.8) \quad \frac{\partial \mathcal{L}}{\partial z} = q(\ \dot{x} \ \partial_z A_x + \dot{y} \ \partial_z A_y + \dot{z} \ \partial_z A_z)$$

Equating both sides,

$$(3B.9) \quad m \ \ddot{z} + q \frac{dA_z}{dt} = q(\ \dot{x} \ \partial_z A_x + \dot{y} \ \partial_z A_y + \dot{z} \ \partial_z A_z)$$

Reworking the second term on the left, dA_z /dt, which is a total derivative,

$$(3B.10) \quad \frac{dA_z}{dt} = \frac{\partial A_z}{\partial x}\frac{\partial x}{\partial t} + \frac{\partial A_z}{\partial y}\frac{\partial y}{\partial t} + \frac{\partial A_z}{\partial z}\frac{\partial z}{\partial t}$$

$$= \partial_x A_z \ \dot{x} + \partial_y A_z \ \dot{y} + \partial_z A_z \ \dot{z}$$

Substituting into 3B.9, and collecting similar terms,

$$(3B.11)\ m\ \ddot{z}\ =\ q[\ \dot{x}\ (\partial_z A_x - \partial_x A_z) + \dot{y}(\partial_z A_y - \partial_y A_z)]$$
$$ma_z = q\ [\ \dot{x}\ [\nabla \times A]_y - \dot{y}[\nabla \times A]_x\ ]\ \text{(equation D4.4c)}$$
$$= q\ [\ \dot{x}\ B_y - \dot{y}\ B_x\ ]\ \text{(equation 3B.4)}$$
$$= q\ [\ v \times B\ ]_z$$

Or for all three components,

$$(3B.12)\ ma = qv \times B \quad \rightarrow F = qv \times B$$

Which we set out to check (see equation 3B.3). Notice that the vector potential A we selected in equation 3B.1 is not unique. For instance, let $A \rightarrow A + \nabla f$, where f is an arbitrary function, we get $B = \nabla \times A \rightarrow \nabla \times (A + \nabla f) = \nabla \times A$, since $\nabla \times \nabla f = 0$ (see appendix D, equation DA.4b).

Let us now work out the Hamiltonian. Reproducing equation 3B.6, except we take the x-component this time,

$$(3B.13)\ P_x = \frac{\partial \mathcal{L}}{\partial \dot{x}} = m\ \dot{x} + qA_x$$

There are similar terms for each of the other components y and z. Repeating equation 2B.1, and then expanding the sum,

$$(3B.14)\ H = \Sigma\ v_i\ p_i - \mathcal{L}(v,q)$$

$$= \dot{x}\ P_x + \dot{y}\ P_y + \dot{z}\ P_z - \mathcal{L}(v,q)$$

We can substitute equation 3B.13 for P_x and for each of the other components.

$$= \dot{x}\,(m\,\dot{x} + qA_x) \; + \; \dot{y}\,(m\,\dot{y} + qA_y)$$
$$\qquad\qquad + \; \dot{z}\,(m\,\dot{z} + qA_z) - \mathcal{L}(v,q)$$
$$= \; m\,\dot{x}^2 + q\dot{x}A_x \; + \; m\,\dot{y}^2 + q\dot{y}A_y$$
$$\qquad\qquad + \; m\,\dot{z}^2 + q\dot{z}A_z - \mathcal{L}(v,q)$$
$$= m\,v^2 + q\,\mathbf{v}\cdot\mathbf{A} - \mathcal{L}(v,q)$$

Substituting equation 3B.1 for $\mathcal{L}$ into the above,

$$= m\,v^2 + q\,\mathbf{v}\cdot\mathbf{A} - (\tfrac{1}{2}\,mv^2 + q\,\mathbf{v}\cdot\mathbf{A})$$
$$= \tfrac{1}{2}\,mv^2$$

We can rewrite this as : $H = \tfrac{1}{2}\dfrac{m}{m}(mv^2) = p^2/2m$

Which is the expected kinetic energy for a charged particle moving in a magnetic field.

Note that $H = T + V$ (see equation 2B.2). In this case, the potential energy $V = 0$.

Though the magnetic force will deflect charged particle, since it is perpendicular to the motion, it does no work on it (Fig. 3A.5).

3C. The Electromagnetic Field

To our previous situation we add an electric field **E**. Equation 3B.3 now reads,

(3C.1) $\mathbf{F} = m\mathbf{a} = q\mathbf{E} + q\mathbf{v} \times \mathbf{B}$

Where the magnetic field **B** acts perpendicular to the plane, and the electric field **E** lies in the plane. This force law is called the Lorentz Law. As previously, associated to the magnetic field there is a vector potential **A**, such that,

(3C.2) $\mathbf{B} = \nabla \times \mathbf{A}$

Also associated to the electric field there is an electric potential,

(3C.3) $\mathbf{E} = -\nabla V$

Where V is measured in the practical unit of volts.

Since the action is taking place in the plane of the paper, our problem is 2-dimensional. Following 3B.2, we can write,

(3C.4) $S = \int [\frac{1}{2} m(\dot{x}^2 + \dot{y}^2) - qV(x,y)] \, dt + q \int \mathbf{A} \cdot \mathbf{dx}$

Note here that $\mathbf{x} = x_i = (x,y)$, is the spatial coordinate, in this case in 2 dimensions. Anticipating Relativity, we'll combine the V term with the **A** term.

(3C.5) $S = \int [\frac{1}{2} m(\dot{x}^2 + \dot{y}^2)] \, dt + q \int [A_i(x_i)dx_i - V(x,y) \, dt]$

Multiply the A_i term by dt/dt,

$$\int A_i(x_i)dx_i \, \frac{dt}{dt} = \int A_i(x_i)v_i \, dt$$

So that equation 3C.5 becomes,

(3C.6) $S = \int [\frac{1}{2} m(\dot{x}^2 + \dot{y}^2)+ q (A_i v_i - V)] \, dt$

Where the Lagrangian is,

(3C.7) $\mathcal{L} = \frac{1}{2} m(\dot{x}^2 + \dot{y}^2)+ q (A_i v_i - V)$

Let's verify if the Lagrangian in equation 3C.7 is the correct form by calculating the equations of motion. We'll concentrate on the x-component of the equation of the Euler-Lagrange equations 2A.9.

(3C.8) $\dfrac{d}{dt}\left(\dfrac{\partial \mathcal{L}}{\partial \dot{x}}\right) = \dfrac{\partial \mathcal{L}}{\partial x}$

$\rightarrow P_x = \dfrac{\partial \mathcal{L}}{\partial \dot{x}} = m\,\dot{x} + qA_x$

Equation 3C.8 becomes,

(3C.9) $\dfrac{d}{dt}(m\,\dot{x} + qA_x) = \dfrac{\partial}{\partial x}[\tfrac{1}{2}\,m\,\dot{x}^2 + q\,(A_i v_i - V)]$

First working the left-hand side,

LHS $\rightarrow \dfrac{d}{dt}(m\,\dot{x} + qA_x) = m\,\ddot{x} + q\dfrac{dA_x}{dt}$

Recall in 2-D, $\dfrac{dA_x}{dt} = \dfrac{\partial A_x}{\partial t}\dfrac{dx}{dt} + \dfrac{\partial A_y}{\partial t}\dfrac{dy}{dt} \equiv \partial_x A_x\,\dot{x} + \partial_y A_x\,\dot{y}$

See appendix E, equation EA.18. Therefore,

LHS $\rightarrow \dfrac{d}{dt}(m\,\dot{x} + qA_x) = \ = m\,\ddot{x} + q\,\partial_x A_x\,\dot{x} + q\,\partial_y A_x\,\dot{y}$

The right-hand side is,

RHS $\rightarrow \dfrac{\partial}{\partial x}[\tfrac{1}{2}\,m\,\dot{x}^2 + q\,(A_i v_i - V)]$

$= \dfrac{\partial}{\partial x}[\tfrac{1}{2}\,m\,\dot{x}^2 + q\,A_x\dot{x} + q\,A_y\dot{y} - qV)]$

$= q\,\partial_x A_x\,\dot{x} + q\,\partial_x A_y\,\dot{y} - q\,\partial_x V$

Equating LHS with RHS,

$m\,\ddot{x} + q\,\partial_x A_x\,\dot{x} + q\,\partial_y A_x\,\dot{y} = q\,\partial_x A_x\,\dot{x} + q\,\partial_x A_y\,\dot{y} - q\,\partial_x V$

Cancelling the common term on both sides,

$$m \ddot{x} + q \partial_y A_x \dot{y} = q \partial_y A_y \dot{y} - q \partial_x V$$

$$m \ddot{x} = q(\partial_x A_y - \partial_y A_x) \dot{y} - q \partial_x V$$

or $\quad\quad\quad ma_x = q E_x + q \dot{y} B_z$

The last step was obtained using equations 3C.2-3. We've obtained the x-component of the general equation (3C.1), which is,

(3C.10) $ma \equiv F = qE + qv \times B$

3D. Gauge Invariance

As it was pointed out earlier, the choice of A in 3B.4, or 3C.2, is not unique. For convenience, it is repeated below.

(3D.1) $B = \nabla \times A$

Let's explore why this is so. A great deal has to do with the curl of a vector. Suppose we make the following transformation,

(3D.2) $A \rightarrow A + \nabla \lambda(x) \equiv A_i + \dfrac{\partial \lambda(x)}{\partial x_i} \equiv A_i + \partial_i \lambda (x)$

Where $\lambda(x)$ is some arbitrary function of position. Calculating the z-component of equation 3D.1,

(3D.3) $[\nabla \times A]_z = \partial_x A_y - \partial_y A_x$

Using equation 3D.2, substitute for each component of A_i in the above equation,

$[\nabla \times A]_z = \partial_x [A_y + \partial_y \lambda (x)] - \partial_y[A_x + \partial_x \lambda (x)]$

$$= \partial_x A_y + \partial_x \partial_y \lambda(x) - \partial_y A_x - \partial_y \partial_x \lambda(x)$$
$$= \partial_x A_y - \partial_y A_x$$

We see that the curl of **A** is left unchanged. This transformation (equation 3D.2) is called a gauge transformation, and if it leaves an equation unchanged, as in this case (equation 3D.3), and in retrospect equation 3C.1, (the Lorentz Law). That equation is said to have gauge invariance.

3E. Gauge Transformation

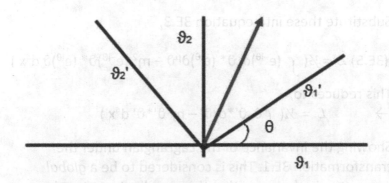

ϑ_2

ϑ_2'

ϑ_1'

θ

ϑ_1

Fig. 3E.1

We are going to explore the concept of gauge invariance in more detail. Consider a scalar field that transforms as:
(3E.1) $\vartheta \rightarrow \vartheta e^{i\theta}$

Where θ is just a number. This is equivalent of rotating the axis by an angle θ.

(3E.2a) $\vartheta' = \vartheta e^{i\theta}$
(3E.2b) $\vartheta'^* = \vartheta e^{-i\theta}$

Consider the following Lagrange (in section 10A, this Lagrangian yields the Klein-Gordon equation 10A.10a):

(3E.3) $\mathcal{L} = \frac{1}{2}\left(\int \partial_\mu \vartheta^* \partial^\mu \vartheta - m^2 \vartheta^* \vartheta \right) d^4x$

If we take the derivatives of equations 3E.2, for instance with respect to x^μ,

(3E.4a) $\dfrac{\partial \vartheta'}{\partial x^\mu} = e^{i\theta} \dfrac{\partial \vartheta}{\partial x^\mu} \equiv e^{i\theta} \partial_\mu \vartheta$

(3E.4b) $\dfrac{\partial \vartheta'^*}{\partial x^\mu} = e^{-i\theta} \dfrac{\partial \vartheta^*}{\partial x^\mu} \equiv e^{-i\theta} \partial_\mu \vartheta^*$

Substitute these into equation 3E.3,

(3E.5) $\mathcal{L} = \frac{1}{2}\left(\int (e^{-i\theta})\partial_\mu \vartheta^* \, (e^{i\theta})\partial^\mu \vartheta - m^2 (e^{-i\theta})\vartheta^* \, (e^{i\theta})\vartheta \; d^4x \right)$

This reduces to,

$\rightarrow \qquad \mathcal{L} = \frac{1}{2}\left(\int \partial_\mu \vartheta'^* \partial^\mu \vartheta' - m^2 \vartheta'^* \vartheta' \; d^4x \right)$

Showing the invariance of the Lagrangian under the transformation 3E.1. This is considered to be a *global* symmetry, in the sense that the coordinate system is rotated rigidly everywhere in space, equally at each and every single point in space. But what if θ varies from point to point. A stronger requirement would be a gauge invariance of the field even though θ varies from point to point.

(3E.6) $\vartheta \rightarrow \vartheta e^{i\theta(x)}$

This is called a local gauge transformation. The changes in our equations would be,

(3E.7a) $\vartheta' = \vartheta\, e^{i\theta(x)}$

(3E.7b) $\vartheta'^* = \vartheta^*\, e^{-i\theta(x)}$

Again take the derivatives,

$$\text{(3E.8a)}\quad \frac{\partial \vartheta'}{\partial x^\mu} = e^{i\theta(x)}\frac{\partial \vartheta}{\partial x^\mu} + i e^{i\theta(x)}\frac{\partial \theta}{\partial x^\mu}\vartheta$$

$$\text{(3E.8b)}\quad \frac{\partial \vartheta'^*}{\partial x^\mu} = e^{-i\theta(x)}\frac{\partial \vartheta^*}{\partial x^\mu} - i e^{-i\theta(x)}\frac{\partial \theta}{\partial x^\mu}\vartheta^*$$

We see that the derivatives have additional terms. And now we substitute these into equation 3E.3,

$$\text{(3E.9)}\quad \mathcal{L} = \tfrac{1}{2}\int d^4x\, [\, (e^{-i\theta(x)})(\partial_\mu\vartheta^* - i\,\partial_\mu\theta\,\vartheta^*)$$

$$X(e^{i\theta(x)})\,(\partial^\mu\vartheta + i\partial^\mu\theta\,\vartheta) - m^2\vartheta^*\,e^{-i\theta(x)}\,\vartheta\,e^{i\theta(x)}\,]$$

$$= \tfrac{1}{2}\int d^4x\, [(\,\partial_\mu\vartheta^* - i\partial_\mu\theta\,\vartheta^*\,)\,(\partial^\mu\vartheta + i\partial^\mu\theta\,\vartheta) - m^2\vartheta]$$

$$= \tfrac{1}{2}\int d^4x\, [\, \partial_\mu\vartheta^*\partial^\mu\vartheta + \partial_\mu\theta\,\partial^\mu\theta\,(\vartheta^*\vartheta)$$

$$+ i\partial^\mu\theta\,\vartheta\,\partial_\mu\vartheta^* - \vartheta^*\partial_\mu\vartheta] - m^2\vartheta^*\,\vartheta$$

Comparing this with equation 3E.5, we see that the Lagrangian is no longer invariant under a gauge transformation as it contains extra terms. But how can we modify this equation so that gauge invariance is restored? The road to take is to look into a different kind of fields. So far, the field ϑ under our consideration is a scalar field. But most fields are as in the case of electric or magnetic field, vector fields, the subject of our next topic.

3F. Transformation of Vector Fields

Vector fields are characterized by one index, as opposed to scalar field with no index. The magnetic field constitutes a 3-vector field. But if we want our field to be invariant under Lorentz transformation then we must compose a 4-vector. So what we need is to tack on a fourth component. Going back to our field theory (section 2C), we denote our vector field as A_μ. The transformations of such a field are going to be constructed in the following manner

(3F.1a) $\vartheta' = \vartheta\, e^{i\theta(x)}$

(3F.1b) $A'_\mu = A_\mu - \dfrac{1}{e}\partial_\mu\theta$

(3F.1c) $\partial_\mu \rightarrow \partial_\mu + ie\, A'_\mu$

Where the constant e is the charge associated with the field. Recall our Lagrangian (equation 3E.3), reproduced below where we drop the ½-factor for convenience,

(3F.2) $\mathcal{L} = \int d^4x\, (\partial_\mu\vartheta'^*\partial^\mu\vartheta' - m^2\vartheta'^*\,\vartheta')$

Putting in place the transformations 3F.1, the first term in the integral will transform as,

$\partial_\mu(\vartheta')^*\partial^\mu\vartheta' = (\partial_\mu\,[\vartheta\, e^{i\theta(x)}) + ie\, A'_\mu\,\vartheta\, e^{i\theta(x)}\,]^*$
$\qquad\qquad X\,(\partial^\mu\,\vartheta\, e^{i\theta(x)} + ie\, A'^\mu\,\vartheta\, e^{i\theta(x)})$

First taking the complement on the right hand-side of the square brackets,

$\qquad = (\partial_\mu\,(\vartheta^*\, e^{-i\theta(x)}) - ie\, A'_\mu\,\vartheta^*\, e^{-i\theta(x)})$
$\qquad\qquad X\,(\partial^\mu\,\vartheta\, e^{i\theta(x)} + ie\, A'^\mu\,\vartheta\, e^{i\theta(x)})$

Taking the derivative and factoring out the exponential function,

$$= e^{-i\theta(x)} (\partial_\mu \vartheta^* - i\vartheta^* \partial_\mu \theta - ie A'_\mu \vartheta^*)$$
$$X\, e^{i\theta(x)} (\partial^\mu \vartheta + i\vartheta \partial^\mu \theta + ie A'^\mu \vartheta)$$

$$= (\partial_\mu \vartheta^* - i\vartheta^* \partial_\mu \theta - ie A'_\mu \vartheta^*)$$
$$X (\partial^\mu \vartheta + i\vartheta \partial^\mu \theta + ie A'^\mu \vartheta)$$

Rearranging,

$$= (\partial_\mu \vartheta^* - i\vartheta^* [\partial_\mu \theta + ie A'_\mu])$$
$$X (\partial^\mu \vartheta + i\vartheta [\partial^\mu \theta + ie A'^\mu])$$

In the square bracket, we have an expression that can be replaced by equation 3F.1b,

$$= (\partial_\mu \vartheta^* - ieA_\mu \vartheta^*) \times (\partial^\mu \vartheta + i eA^\mu \vartheta)$$

Rearranging,

$$= \partial_\mu \vartheta^* \partial^\mu \vartheta + ie(A^\mu \vartheta \partial_\mu \vartheta^* - A_\mu \vartheta^* \partial^\mu \vartheta) + e^2 A_\mu A^\mu \vartheta^* \vartheta$$

We see that the Lagrangian has transformed so far as

$$(3F.3)\mathcal{L} = \int \partial_\mu \vartheta^* \partial^\mu \vartheta + ie(A^\mu \vartheta \partial_\mu \vartheta^* - A_\mu \vartheta^* \partial^\mu \vartheta)$$
$$+ e^2 A_\mu A^\mu \vartheta^* \vartheta - m^2 \vartheta^* \vartheta$$

Though we didn't get to what we wanted to achieve, a Lagrangian that satisfies local gauge invariance, we are much closed when we compared this equation with equation 3E.9. Why are we so intent on pursuing this path? It turns out that the only symmetries with respect to the fundamental forces found in nature are gauge symmetries.

3G. The Electromagnetic Field Tensor

Our object is to construct a Lagrangian when subjected to the principle of least action will gives us the Lorentz law. What we need to do is to bring in the principles so far developed for the construction of a field theory and see if our theory is consistent. There are two basic principles that have guided us so far.

This will be done in chapter 5 in greater details. But for now we mention that the Lorentz invariance assures us that Special Relativity is satisfied, that the speed of light is the same constant in all reference frames, it includes rotational invariance, it guarantees the symmetry of space and time, that equations don't depend where the origin is fixed, or whether there is boost in a given direction. From the point of view of the Lagrangian, this is an easy principle to apply: all that is needed is to assure that the Lagrangian is a scalar, and we do that by checking that the upper indices of the 4-vectors are matched with lower ones

The second principle is gauge invariance. We need to apply all three transformations (equations 3F-1). To insure that the field does not change under a transformation performed at every point in the space in which the physical processes occur, that is, the phase of the complex function ϕ can be changed without altering the outcome of physical experiments, we had to introduce besides the phase factor, also a vector field (3F.1b), along with the covariant derivative (equation 3F.1c).

So we want to derive the Maxwell's equation from the least action principle simply because this principle guarantees that symmetries are associated to conservation laws, such as energy, linear momentum, angular momentum, electric charge, to name a few.

We start out by building a new quantity, a tensor, which has two indices. From our previous result, we must be aware that many of our quantities are derivatives. So the motivation will be to impose the following transformations,

$$(3G.1) \; \partial_v A_\mu \rightarrow \partial_v(A_\mu - \frac{1}{e}\partial_\mu\theta)$$

$$= \partial_v A_\mu - \frac{1}{e}\partial_v\partial_\mu\theta$$

Similarly,

$$(3G.2) \; \partial_\mu A_v \rightarrow \partial_\mu(A_v - \frac{1}{e}\partial_v\theta)$$

$$= \partial_\mu A_v - \frac{1}{e}\partial_\mu\partial_v\theta$$

Subtracting these two equations,

$$(3G.3)$$

$$\partial_v A_\mu - \partial_\mu A_v = (\partial_v A_\mu - \frac{1}{e}\partial_v\partial_\mu\theta) - (\partial_\mu A_v - \frac{1}{e}\partial_\mu\partial_v\theta)$$

$$= \partial_v A_\mu - \partial_\mu A_v \equiv F_{v\mu}$$

This is the *electromagnetic field tensor* - tensor because it contains two indices. Each index can now take 4 values (0,1,2,3). Therefore the number of entries for $F_{\mu v}$ is 16. Four of them are zero, when $\mu = v$, leaving 12 non-zero entries. For example,

(3G.4) $F_{11} = \partial_1 A_1 - \partial_1 A_1 = 0$

Also, $F_{\mu\nu}$ is anti-symmetric, that is, $F_{\mu\nu} = -F_{\nu\mu}$, further reducing number to 6 independent components of the electromagnetic field. For example,

(3G.5) $F_{12} = \partial_1 A_2 - \partial_2 A_1$
$= -(\partial_2 A_1 - \partial_1 A_2)$
$= -F_{21}$

Note that we have three component of the magnetic field, B_1, B_2, B_3 and three components of the electric field, E_1, E_2, E_3, which is exactly what we need to build up the required number of entries for $F_{\mu\nu}$. For example,

(3G.6) $F_{01} = \partial_0 A_1 - \partial_1 A_0$

Recall that the index $\mu = 0$ is just the time component, and $\mu = 1$ is the x-component. Therefore we have,

(3G.7) $F_{tx} = \partial_t A_x - \partial_x A_t$

A good candidate for that component is E_x or E_1. So we make the identification, $F_{01} = E_1$. In all, there are three components with time derivatives, each one being the three components of the electric field. The other three, all involved with spatial derivatives are the three components of the magnetic field. One example is,

(3G.8) $F_{12} = \partial_1 A_2 - \partial_2 A_1$

Switching to x,y, and z convention,

(3G.9) $F_{xy} = \partial_x A_y - \partial_y A_x = B_z$

Switching back,

(3G.10) $F_{12} = B_3$

One complete representation of $F_{\mu\nu}$ is,

(3G.11)

$$F_{\mu\nu} = \begin{pmatrix} F_{00} & F_{01} & F_{02} & F_{03} \\ F_{10} & F_{11} & F_{12} & F_{13} \\ F_{20} & F_{21} & F_{22} & F_{23} \\ F_{30} & F_{31} & F_{32} & F_{33} \end{pmatrix}$$

$$= \begin{pmatrix} 0 & -E_1 & -E_2 & -E_3 \\ E_1 & 0 & B_3 & -B_2 \\ E_2 & -B_3 & 0 & B_1 \\ E_3 & B_2 & -B_1 & 0 \end{pmatrix}$$

To get the other tensor with upper indices, we use,

(3G.12) $F^{\mu\nu} = g^{\mu\delta} g^{\nu\sigma} F_{\delta\sigma}$

Where $g_{\mu\nu} = g^{\mu\nu} = \begin{pmatrix} -1 & 0 & 0 & 0 \\ 0 & 1 & 0 & 0 \\ 0 & 0 & 1 & 0 \\ 0 & 0 & 0 & 1 \end{pmatrix}$ is the metric tensor

(More to say about it in chapter 5).

For example : $F^{01} = g^{0\delta} g^{1\sigma} F_{\delta\sigma}$

The only non-zero terms are $\delta = 0$ and $\sigma = 1$. So

(3G.13) $F^{01} = (-1)(1)(-E_1) = E_1$

Take : $F^{12} = g^{1\delta} g^{2\sigma} F_{\delta\sigma}$

In this case the only non-zero terms are $\delta = 1$ and $\sigma = 2$. So

(3G.14) $F^{12} = g^{11} g^{22} F_{12} = (1)(1) B_3 = B_3$

Going back to construct a Lagrangian for an electromagnetic that is gauge invariant but first, it has to be Lorentz invariant in order to be consistent with Special Relativity (ref. chapter 5). We have to mind our upper and lower indices. We define the Lagrangian density of the electromagnetic field as,

(3G.15) $\mathcal{L} = -\frac{1}{4} F^{\mu\nu} F_{\mu\nu}$

The coefficient $-\frac{1}{4}$ is for convenience. Substitute 3G.3,

(3G.16) $\mathcal{L} = -\frac{1}{4} (\partial^\mu A^\nu - \partial^\nu A^\mu)(\partial_\nu A_\mu - \partial_\mu A_\nu)$

Although this multiplication seems complicated, it simplifies quite easily. Take for example this term,

$$\rightarrow \quad -\frac{1}{4}(F^{0\,1} F_{0\,1} + F^{10} F_{10}) = -\frac{1}{4}[E_1(-E_1) + E_1(-E_1)]$$
$$= \frac{1}{2}(E_1)^2$$
$$= \frac{1}{2} E_x^2$$

We'll get similar terms for the y and z components, which will yield, $\frac{1}{2} E_x^2 + \frac{1}{2} E_y^2 + \frac{1}{2} E_z^2 = \frac{1}{2} E^2$. The spatial components follow a similar pattern.

$$\rightarrow \quad -\frac{1}{4}(F^{12} F_{12} + F^{12} F_{12}) = -\frac{1}{2} B_z^2$$

Again, adding the y and z components yield $-\frac{1}{2} B^2$. The above equation adds up to,

(3G.17) $\mathcal{L} = \frac{1}{2}(E^2 - B^2)$

That is a scalar, and therefore an invariant. We also know that,

(3G.18) $\mathcal{L} = T - V$ (equation 2A.1)

Since both the kinetic energy density and the square of the electric field involve time derivative, we can identify the kinetic energy density as the $\frac{1}{2}E^2$ term, and similarly, $\frac{1}{2}B^2$ as the potential energy density V. We can readily write the energy density, since the Hamiltonian density is,

(3G.19) $H = T + V = \frac{1}{2}(E^2 + B^2)$

3H. The Equation of Motion

The principle of least action means we write down the action and minimize it (see equation 2A.5, reproduced below).

(3H.1) $\delta S = 0$

Consider the Lagrangian in equation 3G.15, reproduced below.

(3H.2) $\mathcal{L} = -\frac{1}{4} F^{\mu\nu} F_{\mu\nu}$

Making the following equivalence for the Euler-Lagrange equations (2C.7),

(3H.3) $\dfrac{d}{dx_\mu} \dfrac{\partial\mathcal{L}}{\partial(\partial A_\mu/\partial x_\nu)} = \dfrac{\partial\mathcal{L}}{\partial A_\nu}$

We start with the LHS.

We do the first integral as follows:

$$\frac{\partial \mathcal{L}}{\partial(\partial A_\mu/\partial x_\nu)} \rightarrow \frac{\partial}{\partial(\partial_\sigma A_\delta)}(-\tfrac{1}{4}(F_{\mu\nu})^2)$$

$$= \frac{\partial F_{\mu\nu}}{\partial(\partial_\sigma A_\delta)}\frac{\partial}{\partial(F_{\mu\nu})}(-\tfrac{1}{4}(F_{\mu\nu})^2)$$

$$= \frac{\partial F_{\mu\nu}}{\partial(\partial_\sigma A_\delta)}(-\tfrac{1}{2} F_{\mu\nu})$$

Substituting equation 3G.3 in the above,

$$\frac{\partial \mathcal{L}}{\partial(\partial A_\mu/\partial x_\nu)} \rightarrow (-\tfrac{1}{2} F_{\mu\nu})\frac{\partial}{\partial(\partial_\sigma A_\delta)}[\partial_\mu A_\nu - \partial_\nu A_\mu]$$

$$= (-\tfrac{1}{2} F_{\mu\nu})[\frac{\partial}{\partial(\partial_\sigma A_\delta)}(\partial_\mu A_\nu)] - (-\tfrac{1}{2} F_{\mu\nu})[\frac{\partial}{\partial(\partial_\sigma A_\delta)}(\partial_\nu A_\mu)]$$

The first term gives for $\sigma = \mu$, $\delta = \nu$: $(-\tfrac{1}{2} F_{\mu\nu})$
The second term gives for $\sigma = \nu$, $\delta = \mu$: $(\tfrac{1}{2}F_{\nu\mu})$

However $F_{\mu\nu}$ is anti-symmetric (equation 3G.5). So this gives a total of $(-\tfrac{1}{2} F_{\mu\nu}) + (-\tfrac{1}{2} F_{\mu\nu}) = -F_{\mu\nu}$

The RHS of equation 3H.3 is zero since the Lagrangian does not depend on A_ν. Therefore,

$$(3H.4) \quad \frac{d}{dx_\mu} F^{\mu\nu} = 0$$

These represent the Maxwell's equations using 4-vector notation. Let's verify one example, $\nu = 0$. First write out the above since it is a summation in μ,

(3H.5) $\partial_0 F^{0\nu} + \partial_1 F^{1\nu} + \partial_2 F^{2\nu} + \partial_3 F^{3\nu} = 0$

Now setting $\nu = 0$

(3H.6) $\partial_0 F^{00} + \partial_1 F^{10} + \partial_2 F^{20} + \partial_3 F^{30} = 0 + \partial_x E_x + \partial_y E_y + \partial_z E_z$

$$= \nabla \cdot E$$

In the absence of charges, this is Maxwell's First Law, (equation 3A.1),

(3H.7) $\nabla \cdot E = 0$.

Another example is $\nu = 1$,

$$\partial_0 F^{01} + \partial_1 F^{11} + \partial_2 F^{21} + \partial_3 F^{31} = -\partial_t E_x + 0 + \partial_y B_z - \partial_z B_y$$

or $$\partial_t E_x = \partial_y B_z - \partial_z B_y$$

This is just the x-component of $\dot{E} = \nabla \times B$, Maxwell's Fourth Law with no current present (equation 3A.4).

3I. Interactions

So far so good. What we've left out is the current density and how it couples to the electromagnet field. There must be terms in the Lagrangian that should involve the current density j_μ. If we look at Maxwell's first equation it contains the time coordinate of the current density, $\mu = 0$,

(3I.1) $\nabla \cdot E = \rho = j_0$

And the fourth law contains the three spatial components of the current density,

(31.2) $\nabla \times B = \dot{E} + j$

In the case that there are charges in spacetime, and subsequently currents if we choose a frame of reference that has undergone a boost, then we must find out how to modify our Lagrangian in order to account for this situation. Now Lorentz invariance requires that any new term in the Lagrangian should be of the type that yields a scalar – in the old language, a dot product. The other suitable candidate is none other than A_μ. Minding our indices, this suggests that our new term for the Lagrangian as $j^\mu A_\mu$. What remains to decide is, will this new term be gauge invariant? According to equation 3F.1b,

(31.3) $j^\mu A_\mu \rightarrow j^\mu (A_\mu + \frac{1}{e} \partial_\mu \varepsilon)$

At first view, it seems that we ran out of luck, as it is not invariant. But the real issue is whether the action is gauge invariant, that is $\delta A = 0$.
So,

(31.4) $\delta j^\mu A_\mu = \frac{1}{e} j^\mu \partial_\mu \varepsilon$

In calculating the change in the action, this is the added term that will show up,

(31.5) $\frac{1}{e} \int j^\mu \partial_\mu \varepsilon \, d^4x$

We must integrate by parts (see Appendix E, equation EB.4). We take that the endpoints are at $\mp$ infinity, where the charges and currents are zero,

(31.6) $\dfrac{1}{e} \int j^\mu \, \partial_\mu \varepsilon \, d^4x = -\dfrac{1}{e} \int \varepsilon \, \partial_\mu j^\mu \, d^4x$

But by the continuity equation (see 3A.13), $\partial_\mu j^\mu = 0$.
Therefore the change in the action will be zero likewise.
Hence the continuity equation becomes the necessary and
sufficient condition for the new term to be added to
Lagrangian so that the latter remain gauge invariant in the
presence of charges or currents. Our full Lagrangian
density then is,

(31.7) $\mathcal{L} = -\frac{1}{4} F^{\mu\nu} F_{\mu\nu} + j^\mu A_\mu$

Repeating the Euler-Lagrangian (3H.3),

(31.8) $\dfrac{d}{dx_\mu} \dfrac{\partial \mathcal{L}}{\partial(\partial A_\mu/\partial x_\nu)} = \dfrac{\partial \mathcal{L}}{\partial A_\nu}$

For example, in equation 3H.5, we will now have,

(31.9) $\partial_0 F^{0\nu} + \partial_1 F^{1\nu} + \partial_2 F^{2\nu} + \partial_3 F^{3\nu} = \dfrac{\partial j^0 A_0}{\partial A_0}$

Now setting $\nu = 0$. For the left-hand side,

(31.10) $\partial_0 F^{00} + \partial_1 F^{10} + \partial_2 F^{20} + \partial_3 F^{30} = 0 + \partial_x E_x + \partial_y E_y + \partial_z E_z$

This is just equal to $\nabla \cdot \mathbf{E}$.

But now the right-hand side is no longer zero,

(31.11) $\dfrac{\partial j^0 A_0}{\partial A_0} = j^0 = \rho$

Equating the two results, and we get the complete Maxwell's first equation, $\nabla \cdot \mathbf{E} = \rho$. It is just an exercise to see that the other three Maxwell's laws are satisfied with what we've defined and derived. The Maxwell's equation (3H.4) in the presence of charges or currents become,

$$(3I.12) \;\; \frac{d}{dx_\mu} F^{\mu\nu} = j^\nu$$

We have accomplished our task: the Lagrangian in equation 3I.7 through the principle of least action yields the equation of motion (Maxwell's equations of electromagnetism) and is fully gauge invariant.

Chapter 4

Thermodynamics

4A. General Remarks

• A thermodynamic system is a specified macroscopic region of the universe.

• It is limited by boundaries: the space outside those boundaries is known as the surroundings, (also as the environment or the reservoir).

• If exchanges occur between the thermodynamic system and the surroundings, the following convention is: the quantities are negative if they leave the system; positive in the opposite case.

• If matter is able to flow in or out of the system, it is said to be open, otherwise, it is closed.

• The system is characterized by a number of macroscopic parameters of which the most important ones are: pressure (P), volume (V), temperature (T); total internal energy (U), and entropy (S).

• One of those equations is the well-known ideal gas law,

(4A.1) $PV = nRT = Nk_BT$, where n is the number of moles, R is the ideal gas constant, N is the number of gas molecules, and k_B is Boltzmann's constant.

• By Thermodynamic Equilibrium we mean that the parameters remain constant with time: for example in the case of the ideal gas, we have the system described by a function f(p,V,T) =0 where f(p,V,T) $\rightarrow$ $PV - Nk_BT$.

● Unless specified otherwise: upper case P is pressure; lower case p is momentum.

4B. The First Law of Thermodynamics

Consider the mechanical work done on a piston which can compress or expand a gas in a closed system (Fig. 4B.1).

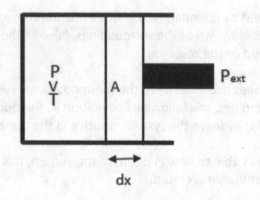

Fig. 4B.1

As depicted, the system is characterized by the parameters P, V and T. The piston, which is part of the environment exerts a pressure P_{ext}. By Newton's 3rd law, the force the gas exerts on the piston is equal and opposite to the force exerted by the piston on the gas. Since both forces are acting on the same area A, we have,

(4B.1) $P_{ext} = - P$

The mechanical work δW is then achieved when the piston is displaced by a certain distance dx.

(4B.2) $\delta W_{ext} = -Fdx = -\frac{F}{A} A \, dx = -PdV$

Or $\quad W_{ext} = - \int PdV$

The internal energy U is the sum of the kinetic and potential energies of all the particles in the system. And Q is the exchanged heat. The first law of thermodynamics states that any work done and any exchange in heat is equal to the internal energy change of the system:

(4B.3) $\Delta U = W + Q$

4C. The Second Law of Thermodynamics

The entropy of a system is defined as,

(4C.1) $\Delta S = \int \frac{\delta Q}{T}$

Where δQ is the heat exchanged of the system, and T is the absolute temperature measured in Kelvin. In statistical mechanics the entropy is also expressed as,

(4C.2) $S = k_B \ln(\Omega)$

Where Ω denotes the number of possible microstates of the system, and $k_B = 1.38 \times 10^{-23} JK^{-1}$ is Boltzmann constant.

If one considers two systems A and B, with entropy S_A and S_B, then the total entropy is just the sum,

(4C.3) $S_{total} = S_A + S_B$

The 2nd law of thermodynamics states that for any number of systems,

(4C.4) $\Delta S \geq 0$

That is, the total entropy of all the systems spontaneously increases, except for adiabatic ($\delta Q = 0$), reversible processes, in which case it is zero.

Example 1: Heat Flow.

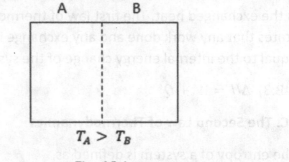

$$T_A > T_B$$

Fig. 4C.1

Consider two closed systems at temperature T_A and T_B, each having a constant volume and separated by a diathermal wall: they are in contact but no work or matter is allowed to be exchanged, except just heat.

(4C.5) $\delta S = \delta S_A + \delta S_B$

$$= \frac{\delta Q_A}{T_A} + \frac{\delta Q_B}{T_B}$$

The combined system is isolated, therefore:

(4C.6) $\delta U = 0 = \delta Q_A + \delta Q_B$ (since W =0)

Or $\rightarrow \delta Q_A = -\delta Q_B$

Substitute into equation 4C.5,

(4C.7) $\delta S = \left(\frac{1}{T_A} - \frac{1}{T_B}\right)\delta Q_A \geq 0$

Note: If $T_A > T_B$, then $\frac{1}{T_A} < \frac{1}{T_B}$ and $\left(\frac{1}{T_A} - \frac{1}{T_B}\right) < 0$. But $\delta S > 0$. That makes $\delta Q_A < 0$. We expect that heat will flow spontaneously from A to B. If system B is at a higher temperature, we can always switch the label A $\leftrightarrow$ B. The net result is that heat always flows spontaneously from a hot body to a cold body.

Example 2: Heat machines

The basic function of a heat machine is to convert heat into work. In principle this requires at least two heat reservoirs, at two different temperatures, say T_H and T_C with $T_H > T_C$.

Assuming a reversible process, over one thermodynamic cycle, the system in its final state recovers its initial state. Applying both laws of thermodynamics (equations 4B.3 and 4C.1),

(4C.8a) $\Delta U = U_F - U_I = 0 \rightarrow W + Q_C + Q_H = 0$

(4C.8b) $\Delta S = 0 = \dfrac{Q_C}{T_C} + \dfrac{Q_H}{T_H}$

Or $\rightarrow \dfrac{Q_C}{Q_H} = -\dfrac{T_C}{T_H}$

The efficiency of a heat machine is defined as the ratio of the amount of produced work (output) to the energy that is supplied (input),

(4C.9) $\eta \equiv \dfrac{output}{input}$

$= \dfrac{|W|}{Q_H}$

$= \dfrac{-W}{Q_H}$

$= \dfrac{Q_C + Q_H}{Q_H}$ (equation 4C.8a)

$= 1 + \dfrac{Q_C}{Q_H}$

Substitute the result in equation 4C.8b,

$$\rightarrow \quad \eta = 1 - \frac{T_C}{T_H}$$

Since $T_H > T_C$, we see that the efficiency, even for a reversible process, is less than 100%.

4D. The Third Law of Thermodynamics

The 3rd law of thermodynamics is stated as: the entropy of a system approaches a constant value as its temperature approaches absolute zero.

Again note that temperature is measured in kelvins.

Suppose we have a crystal at absolute zero (T = 0), with one accessible microstate. From equation 4C.2,

(4D.1) $S_0 = k_B \ln(1) = 0$

Allow the crystal to absorb one photon of energy. All N microstates are possible. Therefore,

(4D.2) $\Delta S = S - S_0 = S = k_B \ln(N)$

Which is a constant.

QED

4E. The Average Kinetic Energy of a Molecule

We make the following assumptions:

(i) A molecule is small compared to the distance between molecules.

(ii) Its interaction with other molecules can be ignored.

(iii) The walls are rigid, and any collision with a molecule will be elastic.

Consider a molecule colliding with a wall in the x-direction. Its momentum will change from mv_x to $-mv_x$.

The change in momentum is then,

(4E.1) $\Delta p_x = mv_x - (-mv_x) = 2mv_x$

The force exerted on the molecule is,

(4E.2) $F = \frac{\Delta p}{\Delta t} = \frac{2mv_x}{\Delta t}$

Since the molecule will move back and forth a distance of $2l$, where l is the distance between two walls. Therefore the time taken is,

(4E.3) $\Delta t = \frac{2l}{v_x}$

Combining these two results, we get

(4E.4) $F = \frac{2mv_x}{\Delta t} = \frac{2mv_x}{2l/v_x} = \frac{mv_x^2}{l}$

This is a force on one molecule. On N molecules,

(4E.5) $F = N\frac{m\bar{v}_x^2}{l}$

Where the bar indicates an average value.

The next step is to consider that we have molecules moving in every directions.

(4E.6) $\bar{v}^2 = \bar{v}_x^2 + \bar{v}_y^2 + \bar{v}_z^2$

Because the molecules are moving with velocities that are random, we can safely say that,

(4E.7) $\bar{v}_x^2 = \bar{v}_y^2 = \bar{v}_z^2$

Therefore,

(4E.8) $\bar{v}^2 = 3\bar{v}_x^2$

$\rightarrow \bar{v}_x^2 = \frac{1}{3}\bar{v}^2$

Substitute this result into equation 4E.5,

(4E.9) $F = N\frac{m\bar{v}^2}{3l}$

The pressure, which is by definition F/A, yields,

(4E.10) $P = \frac{F}{A} = N\frac{m\bar{v}^2}{3Al} = N\frac{m\bar{v}^2}{3V}$

Where the volume is $V = Al$.

Rewriting the above as,

(4E.11) $PV = \frac{1}{3}Nm\bar{v}^2$

Using the formula for an ideal gas, (equation 4A.1),

(4E.12) $PV = Nk_BT = \frac{1}{3}Nm\bar{v}^2$

$\rightarrow m\bar{v}^2 = 3k_BT$

Now we can relate the average kinetic energy of a molecule (= ½ mv²) to the temperature.

(4E.13) $\overline{KE} = \frac{3}{2}k_BT$

Chapter 5
Special Relativity

Recall that so far we have considered velocity to be represented by vectors (Section 1A). At the beginning of the 20th century, a new realization took place that this mathematical language was no longer adequate to describe a new reality. What was that new reality?

Consider the case that a ball is thrown inside a moving car in the same direction of the car. To an observer on the ground, one would add the velocities of the car and the ball. But now consider the ball is replaced by a photon, the particle for light. One can no longer add these two velocities, as the speed of light was reckon to have a constant speed in all inertial frames.

Math is a language. If it does not describe reality, then you need to change the language (math). This was accomplished by the Lorentz transformation equations.

We can describe the order of things (not in the historical sense) as the followings:

(i) There is a new physical reality: the speed of light is the same in all inertial frames;

(ii) Vectors are added in a certain way – call that the normal way.

(iii) That no longer describes the new reality.

(iv) Change the math.

(v) It is accomplished by the Lorentz transformation equations, which gives rise to the 4-vector formalism.

(vi) The bonus point: when $v/c \to 0$ in the Lorentz equations, we get back (ii), the vectors add in the normal way.

So let us begin with the fundamental assumptions of Special Relativity (SR) known as the two postulates.

5A. The Postulates of Special Relativity

Special Relativity is based on two basic postulates:

Postulate 1: The laws of physics are the same in all inertial frames.

Postulate 2: The speed of light is invariant in every inertial frame.

Note 1: The main point for postulate 1 is about observers in different frames. Two observers can choose whatever coordinate system – both chooses the same coordinate system (Cartesian) or different (one, Cartesian; the other, polar) – but what we must keep in mind is that we are dealing with different frames, regardless of the choice of a coordinate system.

Note 2: "Invariant" means that a measurement in one inertial frame will be the same in a different inertial frame, while "constant" means that it doesn't change with respect to time in any frame.

To see this more clearly, consider the 3-D momentum, p_i with i = 1,2,3. In an elastic collision, the total momentum before collision is equal to the total momentum after

collision. The total momentum is said to be a constant —
that is, time independent. In a different frame, a second
observer will also observe that the total momentum
before collision is equal to the total momentum after
collision. But this constant will be different than the
constant observed by the first observer. The 3-D
momentum is not a Lorentz invariant. However, the space-
time interval, which we will define more precisely, is a
Lorentz invariant but it is not a constant quantity. In the
case of the speed of light, it is both a constant (time-
independent) and a Lorentz invariant (the same in every
inertial frame).

5B. The Lorentz Transformation Equations

At the heart of these two postulates are the Lorentz
transformation laws to which we now turn our attention.
Since light moves in a spherical wave front, it can be
described by observer O as a sphere in that inertial frame,

(5B.1) $c^2 t^2 = x^2 + y^2 + z^2$

We can rewrite this as,

(5B.2) $c^2 t^2 - x^2 - y^2 - z^2 = 0$

A second observer O' in a different inertial frame will also
write,

(5B.3) $c^2 t'^2 - x'^2 - y'^2 - z'^2 = 0$

Since these two equation are equal to zero, they are equal
to each other. We have,

(5B.4) $c^2 t^2 - x^2 - y^2 - z^2 = c^2 t'^2 - x'^2 - y'^2 - z'^2$

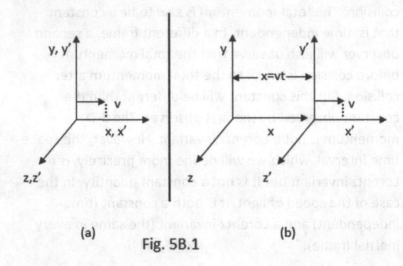

(a)   (b)

Fig. 5B.1

In Fig. 5B.1b, the two frames are at a certain distance away from each other as the primed frame is moving at a speed v going towards the right with respect to the unprimed frame. We can always arrange our two frames such that initially we have y′ = y and z′ = z, (Fig 5B.1a).

Equation 5B.4 then becomes,

(5B.5) $c^2 t'^2 - x'^2 = c^2 t^2 - x^2$

Another assumption is that the transformation laws that relate the two frames O and O′ are linear. Therefore, we can state that,

(5B.6a) $x' = Ax + Bct$

(5B.6b) $ct' = Cx + Dct$

Where A, B, C, and D are to be determined. Those two equations can be put as a single matrix equation such as,

(5B.7) $\begin{bmatrix} x' \\ ct' \end{bmatrix} = \Lambda \begin{bmatrix} x \\ ct \end{bmatrix}$

Where

(5B.8) $\Lambda = \begin{bmatrix} A & B \\ C & D \end{bmatrix}$

(For matrix multiplication, see appendix D, equation DB.5)

Substituting equations 5B.6 into the LHS of equation 5B.5, we get

$\rightarrow c^2 t'^2 - x'^2 = (Cx + Dct)^2 - (Ax + Bct)^2$

Expanding and collecting similar terms we have,

$\rightarrow c^2 t'^2 - x'^2 = (D^2 - B^2)c^2 t^2 - (A^2 - C^2)x^2$

$$+2(DC - AB)xct$$

In order to satisfy the RHS of equation 5B.5, we need

$$D^2 - B^2 = 1$$

$$A^2 - C^2 = 1$$

$$DC = AB$$

This can be satisfied by using the following identity (appendix C, equation CE.5),

(5B.9) $\cosh^2\theta - \sinh^2\theta = 1$

And then we set,

(5B.10) $A = D = \cosh\theta, \ B = C = -\sinh\theta$

Other settings would depict motions different than the one in Fig. 5B.1.

We can now solve for the angle θ. We note that when the two origins are coincident (Fig. 5B.1a), we get the followings: $x' = 0$ and $x = vt$. Substitute that with equations 5B.10 into 5B.6a, we have,

$$\rightarrow x' = 0 = cosh\theta x - sinh\theta\ ct$$

$$= cosh\theta\ vt - sinh\theta\ ct$$

Or

(5B.11) $\tanh \theta = \dfrac{v}{c}$

Using the identity (appendix C, equation CE.6), repeated below,

(5B.12) $\cosh \theta = \dfrac{1}{\sqrt{1-tanh^2\theta}}$

We get

(5B.13) $\cosh \theta = \dfrac{1}{\sqrt{1-\frac{v^2}{c^2}}} \equiv \gamma$; $\sinh \theta = \gamma(\frac{v}{c})$

Note that for massless particle, $v = c$ and $\gamma = 1$; for massive particle, $v < c$ and $\gamma > 1$.

We can now express the Lorentz transformation equations (5B.6a-b) in the familiar form:

(5B.14)

$$x' = \gamma(x - vt);\ ct' = \gamma\left(ct - (\frac{v}{c})x\right);y' = y; z' = z$$

For instance: $x' = Ax + Bct = cosh\theta x - sinh\theta ct$

$$= cosh\theta\left(x - \frac{sinh\theta}{cosh\theta}ct\right) = \gamma(x - \tanh \theta\ ct)$$

$$= \gamma \left(x - \left(\tfrac{v}{c}\right)ct \right) = \gamma(x - vt)$$

Similarly with ct'.

Before we proceed, we must carefully examine equation 5B.5, reproduced below as,

(5B.15) $(c^2 t^2)_{light} = (x^2 + y^2 + z^2)_{light}$

We must reiterate that both sides of the equation referred to the distance squared traveled by light in the unprimed frame. This was done in a 3-D Cartesian coordinate system (also being a flat space) with time as a parameter.

Now an important concept is the space-time interval based on equation 5B.1, which is defined as,

(5B.16) $ds^2 \equiv -c^2 dt^2 + dx_i \, dx^i$

We used differentials since we are going to be dealing with distances, rather than positions. The Einstein summation is applied (see note in appendix D, equation D3.3a), and we set $x = x^1, y = x^2, z = x^3$.

The metric tensor is a 4×4 matrix defined as,

(5B.17) $\eta_{\mu\nu} \equiv \begin{pmatrix} -1 & 0 & 0 & 0 \\ 0 & 1 & 0 & 0 \\ 0 & 0 & 1 & 0 \\ 0 & 0 & 0 & 1 \end{pmatrix}$

With these tools in hand we can rewrite equation 5B.16 as,

(5B.18) $ds^2 = \eta_{\mu\nu} \, dx^\mu \, dx^\nu$

Note 1 – the thumb rule to remember is: no index for a scalar (s); one index for vectors (x^μ); and two indices or more for tensors ($\eta_{\mu\nu}$). The "2" in ds^2 is an exponent, not an index.

Note 2 – the metric tensor is used to lower or raise indices:

(5B.19) $A_\mu = \eta_{\mu\nu} A^\nu$; $A^\mu = \eta^{\mu\nu} A_\nu$ with $\eta_{\mu\nu} = \eta^{\mu\nu}$

Note 3 – the norm of a vector is defined as:

(5B.20) $|A| = (\eta_{\mu\nu} A^\mu A^\nu)^{\frac{1}{2}}$

5C. The Invariance of the Space-Time Interval

The invariance of the interval is so important that it crystalizes everything about SR. No physical theory of the real world can survive without it.

We rewrite equation 5B.7 for differentials with the indices in the right places.

(5C.1) $dx'^\mu = \Lambda^\mu_\nu dx^\nu$

Our assumption that the transformation laws relating the two frames O and O' to be linear demands that the Λ's have an inverse expressed by this restriction:

(5C.2) $\Lambda^\mu_\gamma \Lambda^\nu_\delta \eta_{\mu\nu} = \eta_{\gamma\delta}$

In the primed frame, observer O' writes down her own equation for the interval as,

(5C.3) $ds'^2 = \eta_{\mu\nu} dx'^\mu dx'^\nu$

That the metric tensor is composed of 0's and 1's, and so is invariant. This will no longer be the case in GR as the metric tensor will be replaced by $\eta_{\mu\nu} \rightarrow g_{\mu\nu}$, a quantity to be determined case by case.

Substituting equation 5C.1 into 5C.3, we get

$$(5C.4)\ ds'^2\ =\ \eta_{\mu\nu}\ (\Lambda^\mu_\gamma\ dx^\gamma)(\Lambda^\nu_\delta\ dx^\delta)$$

Using the above restriction (equation 5C.2), we obtain

$$(5C.5)\ ds'^2\ =\ \eta_{\gamma\delta}\ dx^\gamma dx^\delta$$

Now the indices in the Einstein summation are dummy indices, the RHS is just equation 5B.18. Therefore,

$$(5C.6)\ ds'^2\ =\ ds^2$$

We now see the power of the 4-vector formalism. In that mathematical framework, the interval ds is a manifestly invariant quantity: any observer in any initial frame will measure the same space-time interval. The above considerations was for light, which is a massless particle. For a massive particle, moving along a straight line in the x-direction, we have,

$$(5C.7)\ ds^2\ =\ -c^2 dt^2\ +\ dx^2\ <\ 0$$

That's true because no massive particle can travel faster than light, that is, $dx < cdt$.

5D. The Minkowski Space-Time Diagram

The conventional Minkowski diagram is one in which time is no longer a parameter as in Fig. 5D.1, instead it is presented as a coordinate.

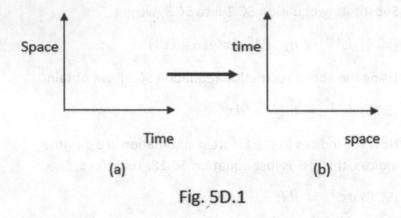

Fig. 5D.1

Note that in Fig. 5D.1a is a graph which is what is most familiar in elementary physics. It is on this graph that velocity is defined as the ratio of distance/time. In SR, that graph is rotated to yield Fig. 5D.1b, and is called a coordinate system. There are some caveats.

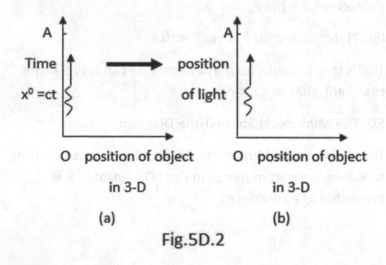

Fig.5D.2

So let's take a very careful look on how one should read the Minkowski diagram (Fig. 5D.2).

A burst of light leaves the origin O and travels through space to a point A along the vertical axis (the word "time" replaced by the position of light). However, the position of an object along the space axis refers to an object placed in a 3-D Cartesian coordinate system, which can be different than light.

There is nothing wrong in making the following mapping for an observer O in some given frame,

$$ct \quad \rightarrow \quad time \quad \rightarrow t$$

Since according to postulate 2, the speed of light in a different frame will be the same (c = c'), an observer O' in a different inertial frame can also make a similar mapping,

$$c't' \quad \rightarrow \quad time \rightarrow t'$$

The speed of light is usually set such that c =1, which is also fine as dimensional units are arbitrary and one can always design a system of units in which c = 1. But most importantly, not to be forgotten is that $x^0 = ct$ measures the distance that light travels in a given time t in that frame. Moreover if the light is emitted from the origin than "ct" is also the position of light at time t in that frame. Time is NOT flowing from position O to position A instead we have a burst of light moving through space from O to A. This means that the axis, $x_0 = ct$, is a vertical spatial axis, no doubt different that the horizontal spatial axis, but nonetheless spatial in kind.

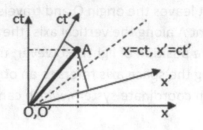

Fig. 5D.3

The other point we want to bring out is why we place the ct-axis orthogonal to the space axis. Perhaps for convenience. But we must be careful not to interpret this as time existing independently outside of space. If we superimpose the primed frame, which is moving away from the unprimed frame at a velocity v, we get Fig. 5D.3. We see that the ct'-x' coordinate system is no longer orthogonal viewed from observer O. Note that both coordinate systems are symmetric about the same line x=t, x'=t', that's because this is the case of the interval being equal to zero in both observers' frames, which is the trajectory of a massless particle. But most importantly the intervals OA and O'A, which have different components in their respective frame, are nevertheless invariant. Both observers will measure the same space-time interval $(OA)^2 = (O'A)^2$.

The Minkowski diagram has its issues. When using it, one must be vigilant and careful as to the claims that can be made. The Minkowski diagram does have its merit – it illustrates quite nicely that the space-time interval is

Lorentz invariant (section 5F) - but one shouldn't be blinded because the math appears to be so appealing.

5E. Time Dilation, Twin Paradox, Length Contraction

To illustrate the time dilation phenomenon, consider again two observers O and O', Fig. 5B.1. In the primed frame, the clock is at rest with respect to O', and necessarily moving with respect to O.

Recall equation 5B.5 reproduced below with some modifications:

(5E.1) $c^2 \Delta \tau'^2 - \Delta x'^2 = c^2 \Delta t^2 - \Delta x^2$

We are using the Δ symbol to indicate that we are looking at intervals of time and position; and the symbol τ for the proper time, which is defined for a clock at rest. From the point of view of O', the clock is at rest, and so we have $\Delta x'^2 = 0$. Therefore,

(5E.2) $c^2 \Delta \tau'^2 = c^2 \Delta t^2 - \Delta x^2$

Divide both sides by c^2, and factor out Δt^2. We get,

(5E.3) $\Delta \tau'^2 = \Delta t^2 (1 - \frac{\Delta x^2}{c^2 \Delta t^2})$

Using $v = \frac{\Delta x}{\Delta t}$ and taking the square-root on both sides, the result is

(5E.4) $\Delta \tau' = \Delta t \left(1 - \frac{v^2}{c^2}\right)^{\frac{1}{2}}$

From equation 5B.13, we have

(5E.5) $\Delta \tau' = \Delta t / \gamma$

With $\gamma > 1$, the proper time $\Delta\tau' < \Delta t$, we say that moving clocks slow down.

In the twin paradox, the frequent question asked is: how do we determine which clock registers the proper time, since a clock can be at rest with respect to one observer but is in motion with respect to another observer.

Consider Alice is in a rocket ship about to undergo a trip to the planet in the nearest star to our sun, say planet Proxima Centauri in the Alpha Centauri star system – about 4.22 light-years. Her twin brother Bob stays back home on planet earth. In this case, the clock in Alice's ship will register the proper time. How do we come to that conclusion? Alice will register two events with one clock: her departure from earth and her arrival on Proxima Centauri. Her clock is at rest with respect to Alice but not with respect to the two events – it has to move from event one (departure) and event two (arrival). On the other hand, Bob will need two clocks to measure the time taken by his sister of these two events: one clock on planet earth, and another one on Proxima Centauri – he needs his good friend Cortney already on Proxima Centauri to record Alice's arrival. So we can see that Bob's clock is NOT the moving clock as it stays back home, registering just one of those two events. Therefore it is Alice who will look younger in age than her twin brother Bob as her clock and everything in her rocket ship will experience this time dilation. Note that the twin paradox is not about traveling to the future but it's all about aging.

This phenomenon has been confirmed in the decay of muons. Comparing the half-life decay between muons producing in the lab with those arriving at very high speed from outer space was a triumph of SR.

In the case of length contraction, we use a car moving at a speed v. Observer O will measure the length L of that car with one clock by registering the time as both front and end of that car will pass in front of him.

(5E.6) $L = v (t_2 - t_1) = v \, \Delta t$

Observer O' is sitting in the car, but in this case, he needs two clocks, one in the front and another in the back of the car, as his car passes through right in front of observer O. He will also measure,

(5E.7) $L' = v (t'_2 - t'_1) = v \, \Delta t'$

We can see that it is observer O who registers the proper time (he uses one clock),

Hence equation 5E.6 becomes,

(5E.8) $L = v (t_2 - t_1) = v \, \Delta t'/\gamma$

Divide equation 5D.8 by 5D.7 yields the result for length contraction,

(5E.9) $L = L'/\gamma$

With γ > 1, L < L'. The moving car (the ruler) is contracted, hence moving rulers contract.

5F. The Space-Time Interval and the Lagrangian

The point that the space-time interval is useful resides on the fact that equation 5B.18, reproduced below, is inadvertently linked to the Lagrangian function.

(5F.1) $ds^2 = \eta_{\mu\nu}\, dx^\mu\, dx^\nu$

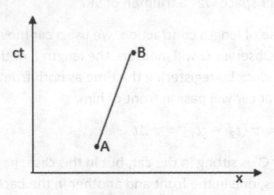

Fig. 5F.1

Where is that link coming from? We want to know how to extremize the path of a free particle (the principle of least action, chapter 2) on a Minkowski diagram.

We write for the proper time, using $d\tau^2 = -ds^2$ and equation 5F.1:

(5F.2) $\tau_{AB} = \int_A^B d\tau = \int_A^B (-\eta_{\mu\nu}\, dx^\mu\, dx^\nu)^{\frac{1}{2}}$

We want to parametrize the line (Fig. 5F.1) such that A = 0 and B = 1, with a parameter σ.

(5F.3) $\tau_{AB} = \int_0^1 (-\eta_{\mu\nu}\, \dfrac{dx^\mu}{d\sigma}\, \dfrac{dx^\nu}{d\sigma})^{\frac{1}{2}}\, d\sigma$

We identify the above equation with the action (equation 2A.4), reproduced below:

(5F.4) $S = \int L(q,\dot{q})\, dt$

With the following mapping:

(5F.5) $S \rightarrow \tau_{AB}\, ;\, L \rightarrow (-\eta_{\mu\nu}\, \dfrac{dx^\mu}{d\sigma}\, \dfrac{dx^\nu}{d\sigma})^{\frac{1}{2}}\, ;\, dt \rightarrow d\sigma$

As we've seen in chapter 2, setting δS = 0 yields the Euler-Lagrange equations (2A.9) (reproduced below),

(5F.6) $\dfrac{\partial L}{\partial x} - \dfrac{d}{d\sigma}\left(\dfrac{\partial L}{\partial \dot{x}}\right) = 0$

Here $x = x^\mu$ and $\dot{x} = dx^\mu/d\sigma$

The first term of equation 5F.6 is,

(5F.7) $\dfrac{\partial L}{\partial x} = \dfrac{\partial}{\partial x}(-\eta_{\mu\nu}\dot{x}^\mu \dot{x}^\nu)^{\frac{1}{2}} = 0$

The second term yields,

(5F.8) $\dfrac{d}{d\sigma}\left(\dfrac{\partial L}{\partial \dot{x}}\right) = \dfrac{d}{d\sigma}\left(\dfrac{\partial}{\partial \dot{x}}(-\eta_{\mu\nu}\dot{x}^\mu \dot{x}^\nu)^{\frac{1}{2}}\right)$

$$= \dfrac{d}{d\sigma}\left(\dfrac{\frac{1}{2}(2)\dot{x}^\mu}{(-\eta_{\mu\nu}\dot{x}^\mu \dot{x}^\nu)^{\frac{1}{2}}}\right) = \dfrac{d}{d\sigma}\left(\dfrac{1}{L}\dfrac{dx^\mu}{d\sigma}\right)$$

Putting the two terms together, we get

(5F.9) $\dfrac{d}{d\sigma}\left(\dfrac{1}{L}\dfrac{dx^\mu}{d\sigma}\right) = 0$

However,

(5F.10) $L = (-\eta_{\mu\nu}\dfrac{dx^\mu}{d\sigma}\dfrac{dx^\nu}{d\sigma})^{\frac{1}{2}} = (\dfrac{d\tau^2}{d\sigma^2})^{\frac{1}{2}} = \dfrac{d\tau}{d\sigma}$

Substitute equation 5F.10 into 5F.9, and multiply throughout by $d\sigma/d\tau$, we get,

(5F.11) $\dfrac{d\sigma}{d\tau}\dfrac{d}{d\sigma}\left(\dfrac{d\sigma}{d\tau}\dfrac{dx^\mu}{d\sigma}\right) = \dfrac{d}{d\tau}\left(\dfrac{dx^\mu}{d\tau}\right) = \dfrac{d^2 x^\mu}{d\tau^2} = 0$

The solution is the equation of a straight line connecting A and B (equation 1A.8), which is what one would expect for a free particle.

5G. Einstein's Famous Equation $E = mc^2$

Einstein derived his equation by looking at events from two observers in two different initial frames.

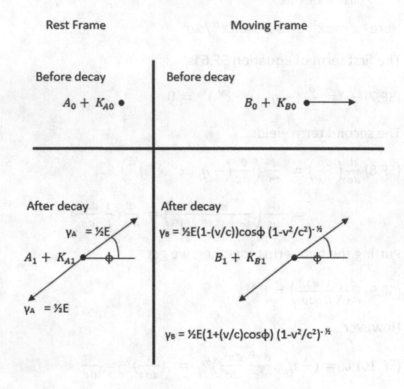

Fig. 5G.1

He knew from experiments previously done that a particle could decay and release gamma rays. He reasoned that when this happened, the particle would lose kinetic energy, and this could only be accounted by a loss of mass. So how did he come to that conclusion? The trick was to analyze the situation both in a rest frame and in a moving frame.

In Fig. 5G.1 the symbol γ refers to the energy of an emitted photon, a value already known by Einstein. The particle has some internal property, which is not needed to be identified, and is labelled as A in the rest frame, and B in the moving frame. The symbol K is for kinetic energy. We use the symbol "0" for the event taking place before the decay, and "1" for after the decay.

By the law of conservation of energy: The energy before decay = the energy after decay. So let us look at the energy in each frame of reference:

(5G.1) In the rest frame:

$$A_0 + K_{A0} = A_1 + K_{A1} + \tfrac{1}{2}E + \tfrac{1}{2}E$$

$$= A_1 + K_{A1} + E$$

(5G.2) In the moving frame:

$$B_0 + K_{B0} = B_1 + K_{B1}$$

$$+ \tfrac{1}{2} E \left(1 - \left(\frac{v}{c}\right) cos\Phi\right) (1 - \frac{v^2}{c^2})^{-\frac{1}{2}}$$

$$+ \tfrac{1}{2} E \left(1 + \left(\frac{v}{c}\right) cos\Phi\right) (1 - \frac{v^2}{c^2})^{-\frac{1}{2}}$$

$$= B_1 + K_{B1} + E(1 - \frac{v^2}{c^2})^{-\frac{1}{2}}$$

Now taking a look at the energy difference of the particle in each of the frame:

(5G.3) In the rest frame (equation 5G.1):

$$\Delta A = A_1 - A_0 = -(K_{A1} - K_{A0}) - E$$

$$= -\Delta K_A - E$$

(5G.4) In the moving frame (equation 5G.2):

$$\Delta B = B_1 - B_0 = -(K_{B1} - K_{B0}) - E(1 - \frac{v^2}{c^2})^{-\frac{1}{2}}$$

$$= -\Delta K_B - E(1 - \frac{v^2}{c^2})^{-\frac{1}{2}}$$

Whether the observer is at rest or moving with respect to the particle, the energy difference (in 5G.3 and 5G.4) should be the same, independent of the internal properties of the decaying particle.

(5G.5) $\Delta A = \Delta B$

$$\rightarrow \quad -\Delta K_A - E = -\Delta K_B - E(1 - \frac{v^2}{c^2})^{-\frac{1}{2}}$$

Calculating the difference in the difference of kinetic energy:

(5G.6) $\Delta K = \Delta K_A - \Delta K_B = E\left(1 - \frac{v^2}{c^2}\right)^{-\frac{1}{2}} - E$

$$= E\left[\left(1 - \frac{v^2}{c^2}\right)^{-\frac{1}{2}} - 1\right]$$

Using $\left(1 - \frac{v^2}{c^2}\right)^{-\frac{1}{2}} \approx 1 + \frac{1}{2}\frac{v^2}{c^2} + \cdots$

$$\Delta K \approx E(1 + \frac{1}{2}\frac{v^2}{c^2} - 1)$$

$$= \frac{1}{2}E\frac{v^2}{c^2}$$

By definition the kinetic energy is,

(5G.7) $K = \frac{1}{2} m v^2$

If the particle was initially at rest, the change in kinetic energy is simply,

(5G.8) $\Delta K = K = \frac{1}{2}mv^2$

Equating 5G.6 and 5G.8, we get

(5G.9) $\frac{1}{2}E\frac{v^2}{c^2} = \frac{1}{2}mv^2$

(5G.10) Therefore, $E = mc^2$

Einstein had reasoned that if the kinetic energy of the particle is smaller by $\frac{1}{2}E\frac{v^2}{c^2}$, the only way this can happen is that the particle must lose mass when emitting radiation, according to equation 5G.10.

Here's an alternative, modern view of deriving the same equation using the 4-vector formalism developed in section 5A.

(5G.11) Again we set c = 1

We will measure the velocity of a free particle with respect to the proper time τ, not the ordinary time t.

Recall equation 5E.5, reproduced below,

(5G.12) $d\tau = dt/\gamma$

And here γ from equation 5B.13 with c = 1,

(5G.13) $\gamma = \frac{1}{\sqrt{1-v^2}}$

The velocity of a free particle with respect to the proper time is defined as,

$(5G.14)$ $u^\beta = \frac{dx^\beta}{d\tau}$

Using the chain rule and equation 5G.12,

$(5G.15)$ $u^\beta = \frac{dt}{d\tau}\frac{dx^\beta}{dt} = \gamma\frac{dx^\beta}{dt}$

Expanding the velocity vector into its 4 components
where k = 1,2,3

$(5G.16)$ $u^\beta = (u^0, u^k) = \gamma\frac{dx^\beta}{dt} = \gamma\left(\frac{dt}{dt}, \frac{dx^k}{dt}\right)$

$\rightarrow$ $u^\beta = \gamma(1, v) = (\gamma, \gamma v)$

The dot product between two 4-vectors (appendix D, equation DC.1) is now defined with the metric tensor of equation 5B.17,

$(5G.17)$ $u^2 = u \cdot u = \eta_{\alpha\beta} u^\alpha u^\beta$

Expanding the above into the temporal and spatial components,

$(5G.18)$ $u^2 = \eta_{00} u^0 u^0 + \eta_{ij} u^i u^j$

Using equation 5G.16,

$(5G.19)$ $u^2 = (-1)\gamma^2 + \gamma^2 v^2 = -1\gamma^2(1 - v^2)$

But squaring equation 5G.13 yields

$(5G.19)$ $\gamma^2 = \frac{1}{1 - v^2}$

Therefore,

$(5G.20)$ $u^2 = u \cdot u = -1$

The 4-vector momentum is defined as mass times velocity, that is,

(5G.21) $p^\alpha = mu^\alpha$

Similarly, the 4-vector momentum has components,

(5G.22) $p^\alpha = (p^0, p^k)$

An important result is to calculate p^2, where again the 2 means squaring, not the component 2. First using equation 5G.20,

(5G.23) $p^2 = mu^\alpha mu^\alpha = m^2 u^2 = -m^2$

Again expanding the above into the temporal and spatial components

$$(5G.24) \; p^2 = \eta_{00} p^0 p^0 + \eta_{ij} p^i p^j$$
$$= -1(p^0)^2 + 1(p^1)^2 + 1(p^2)^2 + 1(p^3)^2$$
$$= -1(p^0)^2 + (\boldsymbol{p})^2$$

Where $\boldsymbol{p}$ is the 3-vector momentum $= (p^1, p^2, p^3)$.

Substituting equation 5G.23 on the LHS:

(5G.25) $-m^2 = -1(p^0)^2 + (\boldsymbol{p})^2$

We define $p^0 \equiv E/c = E$, (using the convention c=1) as the energy of the free particle. Putting this altogether, we get,

(5G.26) $E^2 = (\boldsymbol{p})^2 + m^2$

Putting the factor c back into the equation,

(5G.27) $E^2 = (\boldsymbol{p}c)^2 + m^2 c^4$

For a particle at rest, the momentum is $p = 0$, and so we get $E = mc^2$.

5H. Massless, Massive Particles and Tachyons

Consider now the Lagrangian for a free particle at rest in its own frame. As said above, $p = 0$ and $E = mc^2$. This means that the Lagrangian is

(5H.1) $L = T - V = mc^2$

The action now becomes, (equation 2A.4), reproduced below:

(5H.2) $S = \int L \, dt$

$\qquad = mc^2 \int d\tau$

Using $d\tau = -ds$, we have

(5H.3) $S = -mc^2 \int ds$

The principle of least action demands that,

(5H.4) $\delta S = 0$

As we have mentioned in section 5B, (see below equation5B.13) a massless particle traveling at the speed of light will satisfy 5H.4.

Consider just one spatial axis for simplicity, we have

(5H.5) $ds = ((cdt)^2 - dx^2)^{\frac{1}{2}}$

For $cdt < dx$, the square root is now negative, and we have an imaginary number. We can do the following mapping,

(5H.6) $ds \rightarrow ids$

140

Where i is the imaginary number and ds is now a real number. Then equation 5H.3 becomes,

(5H.7) $S = -imc^2 \int ds$

And this represents a particle with an imaginary mass. Such particles traveling faster than the speed of light are called tachyons in the literature. So far they have not been observed.

·

Chapter 6

General Relativity

One of many considerations that concerned Einstein was that gravity had to affect light. Why? He imagined this thought experiment. Suppose that gravity does **NOT** affect light. Then one could construct the following scheme:

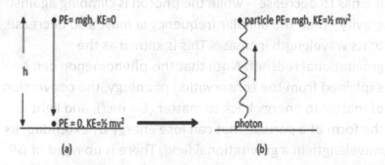

Fig. 6A.1

One could release a particle of mass m at rest at a distance h from the ground (Fig. 6A.1a). As it falls through the gravitational field, its potential energy (PE) is converted to kinetic energy (KE). Einstein knew that this particle could then be converted to a photon (Fig. 6A.1b) from $E = mc^2$, and then one could send the photon to climb against gravity, reconverted to a massive particle after it has climbed the same distance h but now with $KE = \frac{1}{2} mv^2 \neq 0$, if the assumption that gravity does not affect light is correct. Comparing (a) and (b) we see that our particle at height h has gained kinetic energy. One could then repeat this process and create energy out of it. Einstein reasoned that the law of conservation of energy demanded that the photon, a massless particle, must lose energy when climbing up against gravity just like any other massive particle. But how, since light always travels at a constant speed c and a photon has no mass? The only way out was to use what he had already used in his seminal paper on the photoelectric effect,

$$E = \hbar\omega.$$

If E has to decrease – while the photon is climbing against gravity - then the angular frequency ω must also decrease, or its wavelength increase. This is known as the gravitational redshift. Note that this phenomenon can be explained from the conservation of energy, the conversion of matter to energy back to matter (E = mc^2), and light in the form of a particle that can lose energy by extending its wavelength in a gravitational field. There is no need of GR.

6A. Gravitational Time Dilation

In section 5E, we worked out the time dilation in the case of a body moving at constant speed, in which we found that a moving clock slows down. In the Twin Paradox, Alice who travels away will age slower than her stay-at-home twin Bob.

We find a similar situation if our twins are in different gravitational fields.

Suppose our twins are in a rocket which takes off, accelerating against gravity. Bob is in the tail of the rocket (B), while Alice is in the nose of the rocket (A), a distance h from her brother. In terms of the gravitational field, Bob being closer to the center of the earth is in a higher gravitational field than Alice.

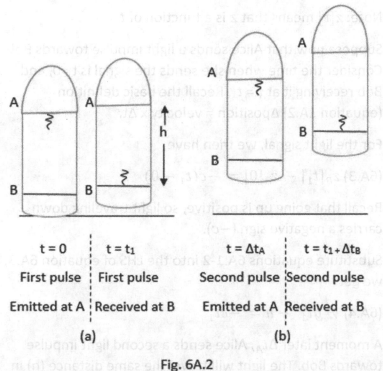

t = 0	t = t_1	t = Δt_A	t = t_1 + Δt_B
First pulse	First pulse	Second pulse	Second pulse
Emitted at A	Received at B	Emitted at A	Received at B

(a)   (b)

Fig. 6A.2

In Fig. 6.2a, Alice sends a pulse at time (t = 0). A moment later (t = t_1), Bob receives this pulse. In Fig. 6.2b, a second pulse is sent by Alice (t = Δt_A). And Bob receives it at a later time (t_1 + Δt_B).

We can describe their two positions using equation 1A.10b as:

For Bob's position,

(6A.1) $z_B[t] = \frac{1}{2} gt^2$

For Alice's position,

(6A.2) $z_A[t] = h + \frac{1}{2} g t^2$

Note: $z[t]$ means that z is a function of t.

Suppose now that Alice sends a light impulse towards Bob. Consider the time when she sends the signal is t =0, and Bob receiving it at $t = t_1$. Recall the basic definition (equation 1A.2) Δposition = velocity x Δt.

For the light signal, we then have,

(6A.3) $z_B[t_1] - z_A[0] = -c(t_1 - 0)$

Recall that going up is positive, so light traveling down carries a negative sign $(-c)$.

Substitute equations 6A.1-2 into the LHS of equation 6A.3, we get

(6A.4) $\frac{1}{2} g t_1{}^2 - h = -ct_1$

A moment later Δt_A, Alice sends a second light impulse towards Bob. The light will travel the same distance (h) in the same time (t_1) and Bob will receive it at some other time, $t_1 + \Delta t_B$. For the second signal we have,

(6A.5) $z_B[t_1 + \Delta t_B] - z_A[\Delta t_A] = -c((t_1 + \Delta t_B) - \Delta t_A)$

Again substitute equations 6A.1-2 this time in the LHS of equation 6A.5, we get

(6A.6) $\frac{1}{2} g(t_1 + \Delta t_B)^2 - (h + \frac{1}{2} g(\Delta t_A)^2)$

$$= -c(t_1 + \Delta t_B - \Delta t_A)$$

Consider the LHS. We square the brackets.

(6A.7) LHS $= \frac{1}{2} g(t_1 + \Delta t_B)^2 - (h + \frac{1}{2} g(\Delta t_A)^2)$

$$= \frac{1}{2} g(t_1^2 + 2t_1\Delta t_B + (\Delta t_B)^2)$$

$$-(h + \frac{1}{2} g(\Delta t_A)^2)$$

Being in a weak gravitational field, we keep only linear terms in Δt_A and Δt_B.

$\rightarrow$ LHS $\approx \frac{1}{2} g t_1^2 + g t_1 \Delta t_B - h$

Equation 6A.6 now reads as,

(6A.8) $\frac{1}{2} g t_1^2 + g t_1 \Delta t_B - h = -c t_1 - c\Delta t_B + c\Delta t_A$

Subtracting equation 6A.4, we get

(6A.9) $g t_1 \Delta t_B = -c\Delta t_B + c\Delta t_A$

Rearranging,

(6A.10) $(g t_1 + c)\Delta t_B = c\Delta t_A$

Now consider that t_1 is the time taken by light to travel the distance between Alice and Bob, which is h. Hence we have $t_1 = h/c$. Substitute that in the above equation, we get,

(6A.11) $\left(g(\frac{h}{c}) + c \right) \Delta t_B = c\Delta t_A$

Divide both sides by c,

(6A.12) $\left(\frac{gh}{c^2} + 1 \right) \Delta t_B = \Delta t_A$

Rearranging,

(6A.13) $\Delta t_B = \dfrac{\Delta t_A}{\left(1 + \frac{gh}{c^2}\right)}$

$$= \Delta t_A \left(1 + \frac{gh}{c^2}\right)^{-1}$$

$$\approx \Delta t_A \left(1 - \frac{gh}{c^2}\right)$$

We see that for Bob in a higher gravitational field his clock (Δt_B) is slower than Alice's.

With respect to an observer on the ground, the Global Position Satellite (GPS) is moving (its clock slows down, an SR effect) and is in a lower gravitational field (its clock speeds up, a GR effect). Both effects must be taken into consideration in order to synchronize the GPS clock with the observer's clock on the ground.

6B. The Equivalence Principle

The Equivalent Principle (EP) states that it is always possible to choose a "locally inertial frame" (LIF) in which gravity is removed. What we have then is a Cartesian frame, which basically allows us to use all the laws of Special Relativity (SR).

So we denote ξ^α as the coordinate system of the free falling frame, and x^μ as any arbitrary coordinate system, often labelled as the "Lab frame". Suppose we have no force acting on a particle inside our free falling frame ($f^\lambda = 0$). We then have no acceleration,

(6B.1) $\frac{d^2\xi^\lambda}{d\tau^2} = 0$ (equation 1C.1)

What happens in the Lab? We can safely say that the ξ^α coordinates are functions of the x^μ coordinate system.

(6B.2) $\xi^\lambda \rightarrow \xi^\lambda(x^\mu)$

Therefore,

(6B.3) $d\xi^\lambda = \dfrac{\partial\xi^\lambda}{\partial x^\mu} dx^\mu$ (appendix E, equation EA.18)

And

(6B.4) $\dfrac{d\xi^\lambda}{d\tau} = \dfrac{\partial\xi^\lambda}{\partial x^\mu} \dfrac{dx^\mu}{d\tau}$

Taking a second derivative (recall 6B.1)

(6B.5) $\dfrac{d^2\xi^\lambda}{d\tau^2} = \dfrac{d}{d\tau}\left(\dfrac{d\xi^\lambda}{d\tau}\right) = \dfrac{d}{d\tau}\left(\dfrac{\partial\xi^\lambda}{\partial x^\mu}\dfrac{dx^\mu}{d\tau}\right) = 0$

Using the Leibniz rule (equation EA.9), we have,

(6B.6) $\dfrac{\partial\xi^\lambda}{\partial x^\mu}\dfrac{d^2 x^\mu}{d\tau^2} + \dfrac{\partial^2\xi^\lambda}{\partial x^\mu \partial x^\nu}\dfrac{dx^\mu}{d\tau}\dfrac{dx^\nu}{d\tau} = 0$

Multiply throughout by $\dfrac{\partial x^\rho}{\partial\xi^\lambda}$, and using the identity

$\dfrac{\partial x^\rho}{\partial x^\mu} = \delta^\rho_\mu$. We get,

(6B.7) $\dfrac{d^2 x^\rho}{d\tau^2} + \Gamma^\rho_{\mu\nu}\dfrac{dx^\mu}{d\tau}\dfrac{dx^\nu}{d\tau} = 0$

Where the Christoffel symbols $\Gamma^\lambda_{\mu\nu}$ are defined as,

(6B.8) $\Gamma^\rho_{\mu\nu} \equiv \dfrac{\partial x^\rho}{\partial\xi^\lambda}\dfrac{\partial^2\xi^\lambda}{\partial x^\mu \partial x^\nu}$

The Christoffel symbols also go by the name of affine connection, and the above equation 6B.7 is also known in the mathematical world as the geodesic equation. So what happens is that the trajectory of a particle in the free falling frame is also the path of a line over a curved manifold.

Note that the Christoffel symbols $\Gamma^{\lambda}_{\mu\nu}$ are functions of both the ξ^{α}'s, the special free falling coordinate system, and the x^{μ}'s, the arbitrary coordinate system.

The next step, which you will find in many textbooks, is to get an equation without the special coordinates ξ^{α}'s. Now we will follow the main line of this argument.

Consider equation 5B.18, reproduced below with the free falling coordinate, after all, the claim according to the EP is that all the laws of SR are valid in this free falling frame.

(6B.9) $ds^2 = \eta_{\alpha\beta}\, d\xi^{\alpha} d\xi^{\beta}$

Substitute equation 6B.3,

(6B.10) $ds^2 = \eta_{\alpha\beta}\, \dfrac{\partial \xi^{\alpha}}{\partial x^{\mu}} dx^{\mu} \dfrac{\partial \xi^{\beta}}{\partial x^{\nu}} dx^{\nu}$

$$= \eta_{\alpha\beta}\, \dfrac{\partial \xi^{\alpha}}{\partial x^{\mu}} \dfrac{\partial \xi^{\beta}}{\partial x^{\nu}} dx^{\mu} dx^{\nu}$$

We write,

(6B.11) $ds^2 = g_{\mu\nu} dx^{\mu} dx^{\nu}$

Where

(6B.12) $g_{\mu\nu} \equiv \eta_{\alpha\beta}\, \dfrac{\partial \xi^{\alpha}}{\partial x^{\mu}} \dfrac{\partial \xi^{\beta}}{\partial x^{\nu}}$

Now the reasoning is to find the relationship between the $\Gamma^{\lambda}_{\mu\nu}$ and $g_{\mu\nu}$, believing that the special ξ^{α} are gone. But as we shall see, the special ξ^{α} are not gone, they are just hidden.

So here we go. Take the derivative of 6B.12, and apply the Leibniz rule (EA.9),

(6B.13) $\frac{\partial g_{\mu\nu}}{\partial x^\lambda} = \eta_{\alpha\beta}\left(\frac{\partial^2 \xi^\alpha}{\partial x^\lambda \partial x^\mu}\right)\frac{\partial \xi^\beta}{\partial x^\nu} + \eta_{\alpha\beta}\frac{\partial \xi^\alpha}{\partial x^\mu}\left(\frac{\partial^2 \xi^\beta}{\partial x^\lambda \partial x^\nu}\right)$

Consider in the first term in the RHS, the factor in the bracket, and using the definition (equation 6B.8):

(6B.14) $\left(\frac{\partial^2 \xi^\alpha}{\partial x^\lambda \partial x^\mu}\right) = \Gamma^\rho_{\lambda\mu}\frac{\partial \xi^\alpha}{\partial x^\rho}$

Similarly for the second term, the factor in the bracket is:

(6B.15) $\left(\frac{\partial^2 \xi^\beta}{\partial x^\lambda \partial x^\nu}\right) = \Gamma^\rho_{\lambda\nu}\frac{\partial \xi^\beta}{\partial x^\rho}$

Substitute both of these in equation 6B.13, we get:

(6B.16) $\frac{\partial g_{\mu\nu}}{\partial x^\lambda} = \eta_{\alpha\beta}\Gamma^\rho_{\lambda\mu}\frac{\partial \xi^\alpha}{\partial x^\rho}\frac{\partial \xi^\beta}{\partial x^\nu} + \eta_{\alpha\beta}\Gamma^\rho_{\lambda\nu}\frac{\partial \xi^\alpha}{\partial x^\mu}\frac{\partial \xi^\beta}{\partial x^\rho}$

Using the definition in equation 6B.12, we have:

(6B.17) $\frac{\partial g_{\mu\nu}}{\partial x^\lambda} = \Gamma^\rho_{\lambda\mu}g_{\rho\nu} + \Gamma^\rho_{\lambda\nu}g_{\mu\rho}$

By permuting the indices, we add two terms and subtract the third one, we get:

(6B.18) $\frac{\partial g_{\mu\nu}}{\partial x^\lambda} + \frac{\partial g_{\lambda\nu}}{\partial x^\mu} - \frac{\partial g_{\mu\lambda}}{\partial x^\nu} = 2\,g_{\rho\nu}\Gamma^\rho_{\lambda\mu}$

We define the inverse of the metric tensor as,

(6B.19) $g^{\mu\sigma}g_{\rho\mu} = \delta^\sigma_\rho$

We get,

(6B.20) $\Gamma^\rho_{\lambda\mu} = \frac{1}{2}\,g^{\nu\rho}\left\{\frac{\partial g_{\mu\nu}}{\partial x^\lambda} + \frac{\partial g_{\lambda\nu}}{\partial x^\mu} - \frac{\partial g_{\mu\lambda}}{\partial x^\nu}\right\}$

Now we did accomplish the task of writing the Christoffel symbols solely in terms of the metric tensor, but hidden in the definition of the metric tensor is the special free falling frame ξ^α, equation 6B.12. This makes the theory still dependent on the free falling frame.

6C. Transformations of Tensors

First let's look on how the affine connections translate from an unprimed coordinate system to a primed one. Starting with an arbitrary vector V^μ,

$$(6C.1) \quad V^{\mu'} = \frac{\partial x^{\mu'}}{\partial x^\mu} V^\mu$$

A convenient rule to remember: If the primed index in the LHS (μ') is the upper index, then the primed index will be also be an upper index on the RHS. Ditto if the primed index in the LHS is the lower index, then it will be also a lower index on the RHS. As always, a repeated index (μ) is a dummy index representing a summation. Secondly, there is a factor for each index of the tensor.

Take the derivative of equation 6C.1 with respect to $x^{\lambda'}$ and using the identity,

$$(6C.2) \quad \frac{\partial}{\partial x'} \rightarrow \frac{\partial x}{\partial x'} \frac{\partial}{\partial x}$$

We get,

$$(6C.3) \quad \frac{\partial V^{\mu'}}{\partial x^{\lambda'}} = \frac{\partial x^\rho}{\partial x^{\lambda'}} \frac{\partial}{\partial x^\rho} \left(\frac{\partial x^{\mu'}}{\partial x^\mu} V^\mu \right)$$

$$= \frac{\partial x^\rho}{\partial x^{\lambda'}} \frac{\partial x^{\mu'}}{\partial x^\mu} \frac{\partial V^\mu}{\partial x^\rho} + \frac{\partial^2 x^{\mu'}}{\partial x^\rho \partial x^\mu} \frac{\partial x^\rho}{\partial x^{\lambda'}} V^\mu$$

The last term is extra. To make this gauge transformation invariant, we must look at how the affine connection will transform (equation 6B.8), reproduced below,

(6C.4) $\quad \Gamma^{\lambda}_{\mu\nu} \equiv \dfrac{\partial x^{\lambda}}{\partial \xi^{\alpha}} \dfrac{\partial^2 \xi^{\alpha}}{\partial x^{\mu} \partial x^{\nu}}$

In the primed one, we have,

(6C.5) $\quad \Gamma^{\lambda'}_{\mu'\nu'} = \dfrac{\partial x^{\lambda'}}{\partial \xi^{\alpha}} \dfrac{\partial^2 \xi^{\alpha}}{\partial x^{\mu'} \partial x^{\nu'}} = \dfrac{\partial x^{\lambda'}}{\partial \xi^{\alpha}} \left(\dfrac{\partial}{\partial x^{\mu'}}\right)\left(\dfrac{\partial}{\partial x^{\nu'}}\right)\xi^{\alpha}$

Apply the identity (6C.2) to each bracket, one at a time,

$$= \dfrac{\partial x^{\lambda'}}{\partial x^{\rho}} \dfrac{\partial x^{\rho}}{\partial \xi^{\alpha}} \left[\left(\dfrac{\partial x^{\tau}}{\partial x^{\mu'}} \dfrac{\partial}{\partial x^{\tau}}\right)\left(\dfrac{\partial x^{\sigma}}{\partial x^{\nu'}} \dfrac{\partial}{\partial x^{\sigma}}\right)\right]\xi^{\alpha}$$

$$= \dfrac{\partial x^{\lambda'}}{\partial x^{\rho}} \dfrac{\partial x^{\rho}}{\partial \xi^{\alpha}} \left[\dfrac{\partial x^{\tau}}{\partial x^{\mu'}} \dfrac{\partial}{\partial x^{\tau}}\left(\dfrac{\partial x^{\sigma}}{\partial x^{\nu'}} \dfrac{\partial \xi^{\alpha}}{\partial x^{\sigma}}\right)\right]$$

And the Leibniz rule,

(6C.6) $\quad \Gamma^{\lambda'}_{\mu'\nu'} = \dfrac{\partial x^{\lambda'}}{\partial x^{\rho}} \dfrac{\partial x^{\rho}}{\partial \xi^{\alpha}} \left[\dfrac{\partial x^{\tau}}{\partial x^{\mu'}} \dfrac{\partial x^{\sigma}}{\partial x^{\nu'}} \dfrac{\partial^2 \xi^{\alpha}}{\partial x^{\tau} \partial x^{\sigma}} + \dfrac{\partial \xi^{\alpha}}{\partial x^{\sigma}} \dfrac{\partial^2 x^{\sigma}}{\partial x^{\mu'} \partial x^{\nu'}}\right]$

Using the definition of the affine connection, equation 6C.4, we get,

(6C.7) $\quad \Gamma^{\lambda'}_{\mu'\nu'} = \dfrac{\partial x^{\lambda'}}{\partial x^{\rho}} \dfrac{\partial x^{\tau}}{\partial x^{\mu'}} \dfrac{\partial x^{\sigma}}{\partial x^{\nu'}} \Gamma^{\rho}_{\tau\sigma} + \dfrac{\partial x^{\lambda'}}{\partial x^{\rho}} \dfrac{\partial^2 x^{\rho}}{\partial x^{\mu'} \partial x^{\nu'}}$

To put the above in standard form we need one more step. Consider the identity,

(6C.8) $\quad \dfrac{\partial x^{\lambda'}}{\partial x^{\rho}} \dfrac{\partial x^{\rho}}{\partial x^{\nu'}} = \delta^{\lambda'}_{\nu'}$

Differentiate both sides with respect to $x^{\mu'}$,

$$\rightarrow \dfrac{\partial}{\partial x^{\mu'}}\left(\dfrac{\partial x^{\lambda'}}{\partial x^{\rho}} \dfrac{\partial x^{\rho}}{\partial x^{\nu'}}\right) = \dfrac{\partial}{\partial x^{\mu'}} \delta^{\lambda'}_{\nu'}$$

The RHS is, $\rightarrow \dfrac{\partial}{\partial x^{\mu'}} \delta^{\lambda'}_{\nu'} = 0$

The LHS is, $\rightarrow \dfrac{\partial x^{\lambda'}}{\partial x^\rho}\dfrac{\partial^2 x^\rho}{\partial x^{\mu'}\partial x^{\nu'}} + \dfrac{\partial x^\rho}{\partial x^{\nu'}}\dfrac{\partial}{\partial x^{\mu'}}\left(\dfrac{\partial x^{\lambda'}}{\partial x^\rho}\right)$

Equating both sides,

(6C.9) $\dfrac{\partial x^{\lambda'}}{\partial x^\rho}\dfrac{\partial^2 x^\rho}{\partial x^{\mu'}\partial x^{\nu'}} + \dfrac{\partial x^\rho}{\partial x^{\nu'}}\dfrac{\partial}{\partial x^{\mu'}}\left(\dfrac{\partial x^{\lambda'}}{\partial x^\rho}\right) = 0$

Therefore,

(6C.10) $\dfrac{\partial x^{\lambda'}}{\partial x^\rho}\dfrac{\partial^2 x^\rho}{\partial x^{\mu'}\partial x^{\nu'}} = - \dfrac{\partial x^\rho}{\partial x^{\nu'}}\dfrac{\partial x^\sigma}{\partial x^{\mu'}}\dfrac{\partial^2 x^{\lambda'}}{\partial x^\rho \partial x^\sigma}$

Substitute for the second term in equation 6C.7,

(6C.11) $\Gamma^{\lambda'}_{\mu'\nu'} = \dfrac{\partial x^{\lambda'}}{\partial x^\rho}\dfrac{\partial x^\tau}{\partial x^{\mu'}}\dfrac{\partial x^\sigma}{\partial x^{\nu'}}\Gamma^\rho_{\tau\sigma} - \dfrac{\partial x^\rho}{\partial x^{\nu'}}\dfrac{\partial x^\sigma}{\partial x^{\mu'}}\dfrac{\partial^2 x^{\lambda'}}{\partial x^\rho \partial x^\sigma}$

The first term is what we would get if the affine connection was a tensor. As we had before in equation 6C.3, the second term is extra. And now we take our cue from the covariant derivative in equation 3F.1c, by adding a term that will cancel the extra term. We define a new covariant derivative. Some textbooks uses the semi-colon convention, others uses the ∇ symbol or an upper case D. That is,

(6C.12a) $V^\mu_{;\nu} \equiv \nabla_\nu V^\mu \equiv D_\nu V^\mu = \dfrac{\partial V^\mu}{\partial x^\nu} + \Gamma^\mu_{\nu\lambda}V^\lambda$

For lower indices,

(6C.12b) $V_{\mu\,;\nu} \equiv \nabla_\nu V_\mu \equiv D_\nu V_\mu = \dfrac{\partial V_\mu}{\partial x^\nu} - \Gamma^\lambda_{\mu\nu}V_\lambda$

For mixed indices,

(6C.12c)

$V^\rho_{\mu\;;\nu} \equiv \nabla_\nu V^\rho_\mu \equiv D_\nu V^\rho_\mu = \dfrac{\partial V^\rho_\mu}{\partial x^\nu} + \Gamma^\rho_{\nu\sigma}V^\sigma_\mu - \Gamma^\lambda_{\nu\mu}V^\sigma_\lambda$

Example: To show that equation 6C.12a transforms as a true tensor.

That is, let $T_{b'}^{a'} = \nabla_{b'} V^{a'}$

If $T_{b'}^{a'}$ is a true tensor, it transforms as

$$T_{b'}^{a'} = \frac{\partial x^b}{\partial x^{b'}} \frac{\partial x^{a'}}{\partial x^a} (T_b^a)$$

Starting with the definition (equation 6C.12a) in the primed coordinate system, we have

(i) $\nabla_{b'} V^{a'} = \dfrac{\partial V^{a'}}{\partial x^{b'}} + \Gamma_{b'i'}^{a'} V^{i'}$

From equation 6C.11,

(ii) $\Gamma_{b'i'}^{a'} = \dfrac{\partial x^{a'}}{\partial x^\rho} \dfrac{\partial x^\tau}{\partial x^{b'}} \dfrac{\partial x^\sigma}{\partial x^{i'}} \Gamma_{\tau\sigma}^\rho - \dfrac{\partial x^\rho}{\partial x^{i'}} \dfrac{\partial x^\sigma}{\partial x^{b'}} \dfrac{\partial^2 x^{a'}}{\partial x^\rho \partial x^\sigma}$

From equation 6C.3,

(iii) $\dfrac{\partial V^{a'}}{\partial x^{b'}} = \dfrac{\partial x^r}{\partial x^{b'}} \dfrac{\partial x^{a'}}{\partial x^a} \dfrac{\partial V^a}{\partial x^r} + \dfrac{\partial^2 x^{a'}}{\partial x^r \partial x^a} \dfrac{\partial x^r}{\partial x^{b'}} V^a$

From equation 6C.1,

(iv) $V^{i'} = \dfrac{\partial x^{i'}}{\partial x^j} V^j$

Substitute ii, iii, iv into i:

(1) $\nabla_{b'} V^{a'} = \dfrac{\partial x^r}{\partial x^{b'}} \dfrac{\partial x^{a'}}{\partial x^a} \dfrac{\partial V^a}{\partial x^r} + \dfrac{\partial^2 x^{a'}}{\partial x^r \partial x^a} \dfrac{\partial x^r}{\partial x^{b'}} V^a$

$$+\left(\frac{\partial x^{a'}}{\partial x^\rho}\frac{\partial x^\tau}{\partial x^{b'}}\frac{\partial x^\sigma}{\partial x^{i'}}\Gamma^\rho_{\tau\sigma} - \frac{\partial x^\rho}{\partial x^{i'}}\frac{\partial x^\sigma}{\partial x^{b'}}\frac{\partial^2 x^{a'}}{\partial x^\rho\partial x^\sigma}\right)\frac{\partial x^{i'}}{\partial x^j}\,V^j$$

Expand the second line:

$$(2)\ \nabla_{b'}V^{a'} = \frac{\partial x^r}{\partial x^{b'}}\frac{\partial x^{a'}}{\partial x^a}\frac{\partial V^a}{\partial x^r} + \frac{\partial^2 x^{a'}}{\partial x^r\partial x^a}\frac{\partial x^r}{\partial x^{b'}}V^a$$

$$+\frac{\partial x^{a'}}{\partial x^\rho}\frac{\partial x^\tau}{\partial x^{b'}}\frac{\partial x^\sigma}{\partial x^{i'}}\Gamma^\rho_{\tau\sigma}\frac{\partial x^{i'}}{\partial x^j}\,V^j - \frac{\partial x^\rho}{\partial x^{i'}}\frac{\partial x^\sigma}{\partial x^{b'}}\frac{\partial^2 x^{a'}}{\partial x^\rho\partial x^\sigma}\frac{\partial x^{i'}}{\partial x^j}\,V^j$$

In the last term, the index j is a dummy, replace it by a; the terms with i' and σ both cancel out top and bottom; the index ρ is also a dummy, replace it by r. We get,

$$(3)\nabla_{b'}V^{a'} = \frac{\partial x^r}{\partial x^{b'}}\frac{\partial x^{a'}}{\partial x^a}\frac{\partial V^a}{\partial x^r} + \frac{\partial^2 x^{a'}}{\partial x^r\partial x^a}\frac{\partial x^r}{\partial x^{b'}}V^a$$

$$+\frac{\partial x^{a'}}{\partial x^\rho}\frac{\partial x^\tau}{\partial x^{b'}}\frac{\partial x^\sigma}{\partial x^{i'}}\Gamma^\rho_{\tau\sigma}\frac{\partial x^{i'}}{\partial x^j}\,V^j - \frac{\partial x^r}{\partial x^{b'}}\frac{\partial^2 x^{a'}}{\partial x^r\partial x^a}V^a$$

The second and last term cancel out.

$$(4)\ \nabla_{b'}V^{a'} = \frac{\partial x^r}{\partial x^{b'}}\frac{\partial x^{a'}}{\partial x^a}\frac{\partial V^a}{\partial x^r} + \frac{\partial x^{a'}}{\partial x^\rho}\frac{\partial x^\tau}{\partial x^{b'}}\frac{\partial x^\sigma}{\partial x^{i'}}\Gamma^\rho_{\tau\sigma}\frac{\partial x^{i'}}{\partial x^j}\,V^j$$

In the first term, the index r is replaced b. In the last term, the index ρ and τ are both dummy indices and are replaced by a and b respectively. Also, in the last term, the index i' cancels out top/bottom, leaving a δ^σ_j, since only σ = j survive, we get

$$(5)T^{a'}_{b'} = \nabla_{b'}V^{a'} = \frac{\partial x^b}{\partial x^{b'}}\frac{\partial x^{a'}}{\partial x^a}\left(\frac{\partial V^a}{\partial x^b} + \Gamma^a_{bj}V^j\right)$$

$$= \frac{\partial x^b}{\partial x^{b'}}\frac{\partial x^{a'}}{\partial x^a}\left(T^a_b\right)$$

QED

Similarly, using the same treatment, equations 6C.12b-c will transform as true tensors.

Reiterating what was shown in this section. We saw that the derivative of a vector (equation 6C.3) and the Christoffel symbols (equation 6C.11) do not transform as tensors. However, the covariant derivative (equations 6C.12abc) does. The importance of this result is that if we have a tensorial equation in one coordinate system, it will transform as a tensor in any other coordinate system.

6D. The Riemann Curvature of Space

(i) Parallel Displacement

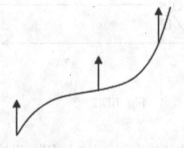

Fig. 6D.1

Parallel displacement means that the vector is transported parallel to itself, and in flat space its components are kept constant as it is moved along a curve.

$$(6D.1) \quad \frac{dV^{\mu}}{ds} = 0$$

If the space is curvilinear, this is no longer tenable. The analogous situation for curved space is that the covariant derivative is zero (equation 6C.12),

(6D.2) $\nabla_s V^\mu = 0 = \dfrac{dV^\mu}{ds} + \Gamma^\mu_{\nu\lambda} V^\nu \dfrac{dx^\lambda}{ds}$

This suggests that,

(6D.3) $dV^\mu = -\Gamma^\mu_{\nu\lambda} V^\nu dx^\lambda$

We can see that the change in the vector V^μ is proportional to several factors, one of which is the change in the coordinate dx^λ. Let us parallel-transport the vector along a full circuit abcd as in Fig. 6D.2.

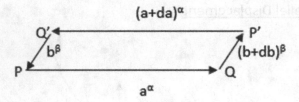

Fig. 6D.2

We expect that the variation in V^μ will be proportional not only to a^α but also to b^β. Since the vectors $(a + da)^\alpha$ and $(b + db)^\beta$ are obtained by parallel exported of a^α and b^β respectively, we have

(6D.4a) $(a + da)^\alpha = a^\alpha - \Gamma^\alpha_{\mu\nu} a^\mu b^\nu$
(6D.4b) $(b + db)^\beta = b^\beta - \Gamma^\beta_{\mu\nu} a^\mu b^\nu$

We now calculate the change in the vector V^μ due to parallel transport from P → Q → P':

(6D.5) $(dV^\mu)_{PQP'} = (dV^\mu)_{PQ} - (dV^\mu)_{QP'}$

$$= -\left(\Gamma^{\mu}_{va}V^{v}\right)_{P}a^{\alpha} - \left(\Gamma^{\mu}_{v\beta}V^{v}\right)_{Q}(b+db)^{\beta}$$

Where the brackets have to be evaluated at points P and Q. We do this by using a Taylor expansion (equation EA.23), in this case around the point P. So the second bracket will break up into two factors $\left(\Gamma^{\mu}_{v\beta}\right)_{Q}$ and $(V^{v})_{Q}$:

(6D.6a) $\left(\Gamma^{\mu}_{v\beta}\right)_{Q} = \left(\Gamma^{\mu}_{v\beta}\right)_{P} + a^{\alpha}\left(\partial_{\alpha}\Gamma^{\mu}_{v\beta}\right)_{P}$

(6D.6b) $(V^{v})_{Q} = (V^{v})_{P} + a^{\alpha}(\partial_{\alpha}V^{v})_{P}$
$\qquad\qquad = (V^{v})_{P} - a^{\alpha}\left(\Gamma^{v}_{\lambda\alpha}V^{\lambda}\right)_{P}$

Where in the second line, we used equation 6D.3 in the last term. With these changes, since all terms in the RHS of equation 6D.5 are around point P, we can drop that subscript. Substitute equations 6D.6a-b, and 6D.4b into 6D.5, we get,

(6D.7) $(dV^{\mu})_{PQP'} = -\Gamma^{\mu}_{va}V^{v}a^{\alpha} - (\Gamma^{\mu}_{v\beta} + a^{\alpha}\partial_{\alpha}\Gamma^{\mu}_{v\beta})$
$$X\ (V^{v} - a^{\alpha}\Gamma^{v}_{\lambda\alpha}V^{\lambda})(b^{\beta} - \Gamma^{\beta}_{\rho\sigma}a^{\rho}b^{\sigma})$$

Keeping only terms of linear order in (ab):

(6D.8) $(dV^{\mu})_{PQP'} = -\Gamma^{\mu}_{va}V^{v}a^{\alpha} - \Gamma^{\mu}_{v\beta}V^{v}b^{\beta}$
$+V^{v}\Gamma^{\mu}_{v\beta}\Gamma^{\beta}_{\rho\sigma}a^{\rho}b^{\sigma} - \partial_{\alpha}\Gamma^{\mu}_{\lambda\beta}V^{\lambda}a^{\alpha}b^{\beta} + \Gamma^{\mu}_{v\beta}\Gamma^{v}_{\lambda\alpha}V^{\lambda}a^{\alpha}b^{\beta}$

Now the change due to parallel transport along the other side, P → Q' → P' can be easily obtained just by switching a ↔ b,

$(6D.9) (dV^\mu)_{PQ'P'} = -\Gamma^\mu_{\nu\alpha} V^\nu b^\alpha - \Gamma^\mu_{\nu\beta} V^\nu a^\beta$

$+V^\nu \Gamma^\mu_{\nu\beta} \Gamma^\beta_{\rho\sigma} b^\rho a^\sigma - \partial_\alpha \Gamma^\mu_{\lambda\beta} V^\lambda b^\alpha a^\beta + \Gamma^\mu_{\nu\beta} \Gamma^\nu_{\lambda\alpha} V^\lambda b^\alpha a^\beta$

The round trip is then:

$(6D.10)\ dV^\mu = (dV^\mu)_{PQ'P'} + (-(dV^\mu)_{PQP'})$

$$\rightarrow dV^\mu = \left[\partial_\alpha \Gamma^\mu_{\lambda\beta} - \partial_\beta \Gamma^\mu_{\lambda\alpha} + \Gamma^\mu_{\nu\alpha} \Gamma^\nu_{\lambda\beta} - \Gamma^\mu_{\nu\beta} \Gamma^\nu_{\lambda\alpha} \right] V^\lambda a^\alpha b^\beta$$

The term in the square bracket is called the Riemann curvature tensor. It is a tensor of rank 4, and involves derivatives of the Christoffel symbols.

$(6D.11)\quad R^\mu_{\beta\alpha\lambda} = \partial_\alpha \Gamma^\mu_{\lambda\beta} - \partial_\beta \Gamma^\mu_{\lambda\alpha} + \Gamma^\mu_{\nu\alpha} \Gamma^\nu_{\lambda\beta} - \Gamma^\mu_{\nu\beta} \Gamma^\nu_{\lambda\alpha}$

Some properties of the Riemann curvature tensor are:

(i) $R^\rho_{\sigma\mu\nu} = g^{\alpha\rho} R_{\alpha\sigma\mu\nu}$

(ii) $R_{\alpha\sigma\mu\nu} = -R_{\sigma\alpha\mu\nu} = -R_{\alpha\sigma\nu\mu} = R_{\mu\nu\alpha\sigma}$

(iii) $R_{\alpha\sigma\mu\nu} + R_{\alpha\nu\sigma\mu} + R_{\alpha\nu\sigma\mu} = 0$

These symmetries show that all 4x4x4x4 components of the Riemann curvature are not independent of each other. In fact, there are only 20 independent components. The dimension of the Riemann curvature tensor is (length) $^{-2}$. In general relativity, mass produces curvature so its scale is (mass/length^3).

(ii) The Ricci Tensor

Contracting two indices of the Riemann tensor, we get the Ricci tensor,

(6D.12) $R_{\mu\nu} = R^{\alpha}{}_{\mu a\nu}$

(iii) The Ricci Scalar

The scalar curvature is obtained by contracting the Ricci tensor,

(6D.13) $R = g^{\mu\nu} R_{\mu\nu}$

6E. Einstein's Field Equations

Let us take a bird's eye view of what the Einstein's field equations are and what do they mean.

Einstein reasoned that if space-time is curved, then the metric tensor in equation 5B.18 must be amended to,

$$\eta_{\mu\nu} \rightarrow g_{\mu\nu}$$

And the space-time interval becomes,

(6E.1) $ds^2 = g_{\mu\nu} dx^{\mu} dx^{\nu}$

Einstein identified (more to say about this in the next section),

(6E.2) $g_{00} \rightarrow - \left(1 + \frac{2\varphi}{c^2}\right)$

Where φ is the gravitational potential, and not the field, given by

(6E.3) $\varphi = - \frac{GM_{source}}{R}$ (equation 1G.5)

Then from Newton's law expressed in terms of Gauss' theorem (equation EC.1),

(6E.4) $\nabla^2 \varphi = 4\pi\rho$

Where ρ is the mass density, Einstein identified,

(6E.5) $T_{00} \rightarrow \rho$

Where T_{00} is the energy-momentum tensor. The rest is history. The Einstein Field equations are then,

(6E.6) $R_{\mu\nu} - \frac{1}{2}g_{\mu\nu}R = \frac{8\pi G}{c^4}T_{\mu\nu}$

Where $R_{\mu\nu}$ is the Ricci tensor (equation 6D.12), and R the Ricci scalar (equation 6D.13).

6F. Weak Gravitational Fields

Often this is referred as the Newtonian limit of GR. Right now, we will demonstrate how Einstein's identification came about (equation 6E.2).

Using equations 6B.7 and 6B.20, rewritten below,

(6F.1) $\frac{d^2x^\lambda}{d\tau^2} + \Gamma^\lambda_{\mu\nu}\frac{dx^\mu}{d\tau}\frac{dx^\nu}{d\tau} = 0$

(6F.2) $\Gamma^\lambda_{\mu\nu} = \frac{1}{2}g^{\lambda\rho}\{\frac{\partial g_{\nu\rho}}{\partial x^\mu} + \frac{\partial g_{\mu\rho}}{\partial x^\nu} - \frac{\partial g_{\mu\nu}}{\partial x^\rho}\}$

Recall the special free falling frame in section 6B. In that frame, we will now consider a particle moving slowly in a "weak stationary gravitational field". Also, we consider the test particle's velocity to be much smaller than the speed of light, v << c. That is for μ, v = 1,2,3, which are the components of the velocity v^i, the second term in equation 6F.1 can be neglected, except for μ, v = 0, we are left with $(x^0 = t)$,

(6F.3) $\dfrac{d^2x^\lambda}{d\tau^2} + \Gamma^\lambda_{00} \dfrac{dx^0}{d\tau} \dfrac{dx^0}{d\tau} = \dfrac{d^2x^\lambda}{d\tau^2} + \Gamma^\lambda_{00} (\dfrac{dt}{d\tau})^2 = 0$

And from equation 6F.2,

(6F.4) $\Gamma^\lambda_{00} = \dfrac{1}{2} g^{\lambda\rho} \left\{ \dfrac{\partial g_{0\rho}}{\partial x^0} + \dfrac{\partial g_{0\rho}}{\partial x^0} - \dfrac{\partial g_{00}}{\partial x^\rho} \right\}$

$= \dfrac{1}{2} g^{\lambda\rho} \left\{ \dfrac{\partial g_{0\rho}}{\partial t} + \dfrac{\partial g_{0\rho}}{\partial t} - \dfrac{\partial g_{00}}{\partial x^\rho} \right\}$

To the observer in that special free falling frame, the gravitational field is static, and so the derivative of the field with respect to time is zero.

Therefore, equation 6F.4 reduces furthermore to,

(6F.5) $\Gamma^\lambda_{00} = -\dfrac{1}{2} g^{\lambda\rho} \dfrac{\partial g_{00}}{\partial x^\rho}$

Because the field is weak, we can write the metric tensor as,

(6F.6) $g_{\lambda\rho} = \eta_{\lambda\rho} + h_{\lambda\rho}, \; |h_{\lambda\rho}| \ll 1$

Where $\eta_{\lambda\rho}$ is the metric tensor in flat space (equation 5B.17).

Thus equation 6F.5 becomes

(6F.7) $\Gamma^\lambda_{00} = -\dfrac{1}{2} \eta^{\lambda\rho} \dfrac{\partial h_{00}}{\partial x^\rho}$

And equation 6F.3 becomes

(6F.8a) $\dfrac{d^2x^i}{d\tau^2} = \dfrac{1}{2} (\dfrac{dt}{d\tau})^2 \dfrac{\partial h_{00}}{\partial x^i}$ for $\lambda, \rho = 1, 2, 3 \equiv i$

(6F.8b) $\frac{d^2t}{d\tau^2} = 0$ for $\lambda, \rho = 0$

The second equation (6F.8b) tells us that dt/dτ is a constant and so we can divide the first equation (6F.8a) by (dt/dτ)^2, we get **(bold-face for vectors)**,

(6F.9) $\frac{d^2\mathbf{x}}{dt^2} = \frac{1}{2} \boldsymbol{\nabla} h_{00}$

The corresponding Newtonian result is,

(6F.10) $\frac{d^2\mathbf{x}}{dt^2} = -\boldsymbol{\nabla}\varphi$ (equation 1E.4)

Where φ is the gravitational potential at a distance r from the center of a body of mass M ($\varphi = \frac{V}{m_{test}}$ in equation 1E.4).

(6F.11) $\varphi = -\frac{GM}{r}$ (equation 1G.5(5))

Equating 6F.9 and 6F.10, we get

(6F.12) $h_{00} = -2\varphi +$ constant

From equation 6F.6, Einstein deduced (recall $\eta_{00} = -1$),

(6F.13) $g_{00} = -\left(1 + \frac{2\varphi}{c^2}\right)$

Where we restored c. And that is equation 6E.2.

6G. Schwarzschild's Solution to the Einstein Field Equations

One rare situation for which there is an exact solution in GR is the case of a perfectly symmetric spherical object.

Consider a sphere (Fig. 6G.1) and a small volume element:

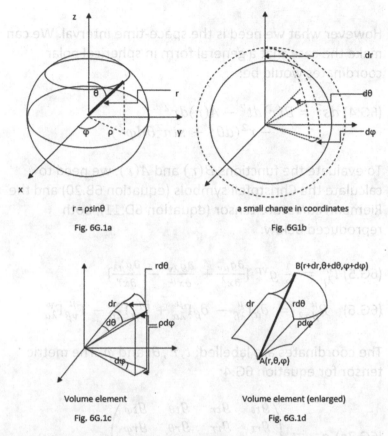

r = ρsinθ

Fig. 6G.1a

a small change in coordinates

Fig. 6G1b

Volume element

Fig. 6G.1c

Volume element (enlarged)

Fig. 6G.1d

The arc forming from a change in the angle θ is,

(6G.1) $ds_\theta = rd\theta$

For the angle φ

(6G.2) $ds_\varphi = \rho d\varphi = r\sin\theta d\varphi$

A small distance between points A and B (Fig. 6G.1d) is

(6G.3) $ds^2 = dr^2 + r^2d\theta^2 + r^2\sin^2\theta d\varphi^2$

However what we need is the space-time interval. We can make the case that a general form in spherical polar coordinates would be:

(6G.4) $ds^2 = B(r)dt^2 - A(r)dr^2$
$$- r^2(d\theta^2 + sin^2\theta d\varphi^2)$$

To evaluate the functions $B(r)$ and $A(r)$, we need to calculate the Christofell symbols (equation 6B.20) and the Riemann curvature tensor (equation 6D.11), both reproduced below:

(6G.5) $\Gamma^\rho_{\lambda\mu} = \frac{1}{2} g^{\nu\rho} \{\frac{\partial g_{\mu\nu}}{\partial x^\lambda} + \frac{\partial g_{\lambda\nu}}{\partial x^\mu} - \frac{\partial g_{\mu\lambda}}{\partial x^\nu}\}$

(6G.6) $R^\mu_{\beta\alpha\lambda} = \partial_\alpha\Gamma^\mu_{\lambda\beta} - \partial_\beta\Gamma^\mu_{\lambda\alpha} + \Gamma^\mu_{\nu\alpha}\Gamma^\nu_{\lambda\beta} - \Gamma^\mu_{\nu\beta}\Gamma^\nu_{\lambda\alpha}$

The coordinates are labelled: t, r, θ and φ. The metric tensor for equation 6G.4:

(6G.7a) $g_{\mu\nu} \equiv \begin{pmatrix} g_{tt} & g_{tr} & g_{t\theta} & g_{t\varphi} \\ g_{rt} & g_{rr} & g_{r\theta} & g_{r\varphi} \\ g_{\theta t} & g_{\theta r} & g_{\theta\theta} & g_{\theta\varphi} \\ g_{\varphi t} & g_{\varphi r} & g_{\varphi\theta} & g_{\varphi\varphi} \end{pmatrix}$

$$= \begin{pmatrix} B(r) & 0 & 0 & 0 \\ 0 & -A(r) & 0 & 0 \\ 0 & 0 & -r^2 d\theta^2 & 0 \\ 0 & 0 & 0 & -r^2 sin^2\theta d\varphi^2 \end{pmatrix}$$

The inverse is,

(6G.7b)

166

$$g^{\mu\nu} =$$

$$\begin{pmatrix} B(r)^{-1} & 0 & 0 & 0 \\ 0 & -A(r)^{-1} & 0 & 0 \\ 0 & 0 & (-r^2 d\theta^2)^{-1} & 0 \\ 0 & 0 & 0 & (-r^2 \sin^2\theta d\varphi^2)^{-1} \end{pmatrix}$$

Evaluating Γ_{tt}^t

(6G.8) $\Gamma_{tt}^t = \frac{1}{2} g^{\nu t} \left\{ \frac{\partial g_{tv}}{\partial x^t} + \frac{\partial g_{tv}}{\partial x^t} - \frac{\partial g_{tt}}{\partial x^v} \right\}$

The only surviving terms are those for which $v = t$, hence

$\rightarrow \Gamma_{tt}^t = \frac{1}{2} g^{tt} \left\{ \frac{\partial g_{tt}}{\partial x^t} + \frac{\partial g_{tt}}{\partial x^t} - \frac{\partial g_{tt}}{\partial x^t} \right\} = 0$

Where we have every term in that bracket is not a function of the time t. A non-zero term is,

(6G.9) $\Gamma_{tr}^t = \frac{1}{2} g^{\nu t} \left\{ \frac{\partial g_{rv}}{\partial x^t} + \frac{\partial g_{tv}}{\partial x^r} - \frac{\partial g_{\mu\lambda}}{\partial x^v} \right\}$

Again the term outside the bracket establishes that $v = t$. The only surviving term is the second term, so $g^{tt} = B(r)^{-1}$, and $\Gamma_{tr}^t = \frac{1}{2} B(r)^{-1} \left(\frac{\partial B(r)}{\partial x^r} \right) = \frac{B'}{2B}$ where the prime indicates a derivative with respect to r.

We will quote the rest of the non-zero Christoffel symbols:

(6G.10) $\Gamma_{rr}^r = \frac{A'}{2A}$; $\Gamma_{\theta\theta}^r = -\frac{2\sin^2\theta}{A}$; $\Gamma_{tt}^r = \frac{B^2}{2A}$

$\Gamma_{r\theta}^\theta = \Gamma_{\varphi r}^\varphi = \frac{1}{r}$; $\Gamma_{\varphi\varphi}^\theta = -\sin\theta\cos\theta$; $\Gamma_{\varphi\theta}^\varphi = \cot\theta$

The next step is to evaluate the Ricci tensor (equation 6D.12). Here we will only quote one non-zero term,

$$(6G.11) \ R_{tt} = - \frac{B''}{2A} + \frac{1}{4}\left(\frac{B'}{A}\right)\left(\frac{A'}{A} + \frac{B'}{B}\right) - \frac{1}{r}\frac{B'}{A}$$

We can argue from the Einstein Field Equation (6E.7), reproduced below:

$$(6G.12) \ R_{\mu\nu} - \frac{1}{2}g_{\mu\nu}R = \frac{8\pi G}{c^4}T_{\mu\nu}$$

In a vacuum, $T_{tt} = 0$, and so both the Ricci tensor and the Ricci scalar are also zero. From equation 6G.11, we get two equations,

$$(6G.13a) \ \frac{A'}{A} + \frac{B'}{B} = 0 \text{ (from the middle term)}$$

$$(6G.13b) \ -\frac{B''}{2A} - \frac{1}{r}\frac{B'}{A} = 0 \text{ (from 1st and 3rd terms)}$$

$$\rightarrow \frac{B''}{2} + \frac{B'}{r} = 0$$

From the last equation, a solution is,

$$(6G.14) \ B(r) = C_1 + \frac{C_2}{r}$$

From $g_{00} = -(1 + 2\varphi)$ (equation 6E.2)

And

$$(6G.15) \ \varphi = -Gm/r \text{ (equation 6F.11)}$$

A good choice is then $C_1 = -1, C_2 = 2Gm$

$$\rightarrow B(r) = 1 - \frac{2Gm}{r}$$

From 6G.13a, we get $A(r) = \dfrac{C_3}{1 - \frac{2Gm}{r}}$

For $r \rightarrow \infty$, $A(r) \rightarrow 1$ for flat space. Therefore choose $C_3 = 1$,

$$\rightarrow A(r) = \frac{1}{1 - \frac{2Gm}{r}}$$

Define $R_{\text{Schwarzschild}} \equiv R_* = 2Gm$, the space-time interval (equation 6G.4) now reads,

$$(6G.16) \quad ds^2 = \left(1 - \frac{R_*}{r}\right) dt^2 - \left(1 - \frac{R_*}{r}\right)^{-1} dr^2$$

$$- r^2(d\theta^2 + \sin^2\theta d\varphi^2)$$

With $r = R_*$, the metric tensor blows up. Note: this singularity is coordinate dependent.

Part 2

QUANTUM MECHANICS

Chapter 7

The Essential Quantum Mechanics

How do we measure the velocity of a car? We see the car because an enormous number of photons are hitting the car in all directions, and some of them will reach our eyes. We can then note where it is at a given time, call that x_1 and t_1. At a later time, we then observe the car again but at x_2 and t_2. A whole set of these points can be plotted, and from there, we can figure out the velocity, and determine if it is in uniform motion or if it is accelerating or decelerating, etc.

But consider the case of an electron.

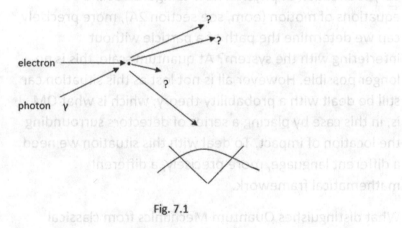

Fig. 7.1

So it goes for an electron, to find out anything about it, the idea is to shoot a whole bunch of photons, in this case short-wave or high-energy photons due to the smallness of the electron. We get lucky as one of those photons hits the electron, and with very much luck, it bounces in the right way to reach our eyes. This is what the photon would be telling us if it could speak, "Sir, that electron is right there," call that position X, even though X is really a smeared area as our electron was jiggling around when it was hit, "but then guess what Sir, I've also thrown it off its position, and I haven't a clue in what direction it's going." Hitting the electron with a photon would be like hitting the car in the previous case with a missile. It would be unlikely that the car would have continued along its original path. It would follow some undetermined path as in Fig. 7.1. Hitting the electron with a second photon to get another position and time, x_2 and t_2, would be a very, very difficult task. In other words, the electron's path after the collision with the photon becomes unpredictable. This strikes at the very foundation of physics: how to figure out the equations of motion (eom, see section 2A), more precisely, can we determine the path of a particle without interfering with the system? At quantum scale, this is no longer possible. However all is not lost as this situation can still be dealt with a probability theory, which is what QM is, in this case by placing a series of detectors surrounding the location of impact. To deal with this situation we need a different language, more precisely, a different mathematical framework.

What distinguishes Quantum Mechanics from classical physic are the concepts of states and observables. For instance, the position x in classical physics plays both roles of indicating a state and an observable. In QM, these two roles are handled differently: the quantum state can be represented by a function $\psi(x)$ or a vector $| x >$, where we use the Dirac notation $| >$, read as a "ket"; while the observables are represented by operators, and as the name implies, these operate on quantum states. They can be just a real or complex number, a variable in the algebraic sense, matrices or derivatives in some cases. The stunning difference is that those operators don't necessarily commute. Hence one of the primary tasks is to establish some commutation relationships between observables and their conjugates - to be defined later on. Needless to say that QM will have a radically different mathematical structure than classical physics.

To begin, we will focus our attention on the wave function, the operative part is "function", as this object is mathematical in nature, and not to be taken as a real wave. If you keep that in mind, a lot of confusion about QM will dissipate.

7A. Mathematical Foundation of Quantum Mechanics

(7A.1) We will use the Dirac notation, that is, the wave function is a vector V denoted by a "ket", $| V >$.

(7A.2) A ket can be multiplied by any scalar α, β... denoted by $\alpha| V >$, $\beta| V >$...

Note we can also write $\alpha| V >$ as $| \alpha V >$, that is, a new vector $| V' > \equiv | \alpha V > \equiv \alpha| V >$.

(7A.3) There is a dual vector, the "bra", denoted by $< V |$.

(7A.4) Note that $< V' | = < \alpha V | = < V | \alpha^*$, where α^* is the complex conjugate of α (see appendix K, equation KA.2).

(7A.5) Vectors can also be represented by matrices. In the

case of the ket vector, | V >, it will be a column matrix, while the bra < V | will be a row matrix, with its elements as the complex conjugate of every element of the vector |V >.

$$|V> = \begin{bmatrix} V_1 \\ V_2 \\ V_3 \\ V_4 \end{bmatrix} \qquad <V| = [V_1^* \quad V_2^* \quad V_3^* \quad V_4^*]$$

(7A.6) Any vector | V > in an n-dimensional Hilbert space can be written as a linear combination of n linearly independent unit vectors | 1 >, | 2 >... | n >.

(7A.7) A set of n linearly independent vectors in an n-dimensional space is called a *basis*, and so we can write,

| V > = Σ V_i | i >,

where i = 1... n, V_i are the components of the vector, and the | i >'s are the basis vectors.

(7A.8) We can construct an inner product between two vectors. The analog of the dot product between two vectors **V** and **W** (see appendix D, equation DA.3b), where **V·W** = VW cosθ, is < V | W >. For instance,

$$<W | V> = [W_1^* \quad W_2^* \quad W_3^* \quad W_4^*] \begin{bmatrix} V_1 \\ V_2 \\ V_3 \\ V_4 \end{bmatrix}$$

$$= W_1^* V_1 + W_2^* V_2 + W_3^* V_3 + W_4^* V_4$$

Note that

(7A.9) $< V \mid W > = < W \mid V >^* = \Sigma \Sigma V_i^* W_j < i \mid j >$.

(7A.10) If the basis vectors have unit length, then they form an *orthonormal* base. From (7A.7) above, we can find the jth component of the vector $\mid V >$ as such,

$\mid V > = \Sigma V_i \mid i >$,

Multiply both sides by $< j \mid$

$< j \mid V > = \Sigma V_i < j \mid i >$
$\qquad = \Sigma V_i \delta_{ij}$ - using (7A.6)
$\qquad = V_j$

An important result with $V_i = < i \mid V >$, equation 7A.7 can now be written as,

$\mid V > = \Sigma \mid i >< i \mid V >$.

(7A.11) An *operator* Ω is an instruction to transform a vector $\mid V >$ into another vector $\mid V' >$. This can be written as,

$\Omega \mid V > = \mid \Omega V > = \mid V' >$.

Note that the operator Ω acts on the *right*.

(7A.12) Operators are said to be *linear* if they obey the following rules: for any scalars α and β,

(7A.12a) $\Omega \alpha \mid V > = \alpha \Omega \mid V >$

(7A.12b) $\Omega \{\alpha \mid V > + \beta \mid V' >\} = \alpha \Omega \mid V > + \beta \Omega \mid V' >$

The action of two successive operators Ω and Λ in general will not be commutative. We designate the commutation relationship as:

(7A.13a) $\Omega\Lambda - \Lambda\Omega \equiv [\Omega,\Lambda]$

Where the commutator is designated by $[\,\bullet, \bullet\,]$.

And anti-commutation relations as:

(7A.13b) $\Omega\Lambda + \Lambda\Omega \equiv \{\Omega,\Lambda\}$

Where the anti-commutator is designated by $\{\,\bullet, \bullet\,\}$.

(7.14) Recall that $|\,V> = \Sigma\,|\,i><i\,|\,V>$. We now define the projector operator as,

$P_i = |\,i><i\,|$

The importance of this operator is that,

$I = \Sigma\,P_i$, where I is the identity operator.

(7A.15) In regard to the dual vector, the bra $<V\,|$, we define the operator acting on it as,

$<V'\,| = <\Omega V\,| = <V\,|\Omega^\dagger$, where $\Omega^\dagger$ is called the adjoint operator.

Note that $\Omega^\dagger$ acts on the *left*.

Also, $<V'\,| = <\alpha V\,| = <V\,|\alpha^*$

Therefore, $<V\,|\Omega^\dagger = <V\,|\alpha^*$

In a given basis, the adjoint operation is the same as taking the transpose conjugate (equation 7A.5).

(7A.16) An important theorem for two operators Ω and Λ is,

$(\Omega\Lambda)^\dagger = \Lambda^\dagger\Omega^\dagger$

(7A.17) Since an observable is represented by an operator we must keep in mind that observables must be real numbers. To get this result we must define what is an Hermitian operator. An operator is said to be Hermitian if,

$H^\dagger = H$

(7A.18) An operator is said to be *unitary* if,

$U^\dagger U = UU^\dagger = I$, where I is the identity operator.

Note that $U^\dagger = U^{-1}$, where U^{-1} is the inverse of U.

A unitary operator preserves the inner product:

Proof: consider $| V' > = U| V >$ and $| W' > = U| W >$

Then $< W' | V' > = < W |U^\dagger U | V > = < W | V >$

(7A.19) An important problem involves the situation when,

$\Omega| V > = \omega| V >$, where ω is a number, real or complex.

We say that the operator Ω rescales the vector $| V >$ by a factor ω. That equation is called an *eigenvalue* equation, and | V > is an *eigenket* of Ω with *eigenvalue* ω.

(7A.20) Consider:

< V |Ω| V > = ω < V | V >, and < V |Ω†| V > = ω* < V | V >. If Ω is hermitian, then Ω = Ω†. Therefore ω = ω*, and ω is real. Since what we measure are observables, and their measurements must be real numbers, Hermitian operators are the perfect candidate for observables in QM.

$$|f_n\rangle \leftrightarrow \begin{bmatrix} f_n(x_1) \\ f_n(x_2) \\ \vdots \\ f_n(x_n) \end{bmatrix}$$

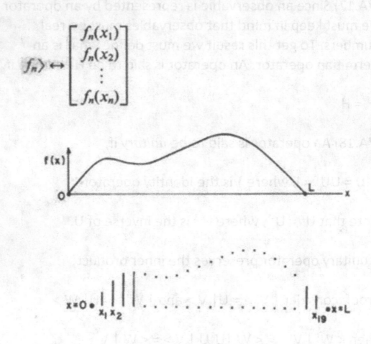

Fig. 7A.1

(7A.21) So far we have dealt with vectors. Now we want to bring in continous functions into our formalism as these will play an important role. We take a function f(x) along a

180

certain interval between 0 and L. Divide this into equal parts, say n=20 parts. Let x=L/20, 2L/20...19L/20 (see Fig.7A.1). We denote the ket $| f_n(x) >$ as the discrete approximation of the continuous function f(x). The basis vectors in this space are then

$$| x_i > = \begin{bmatrix} 0 \\ 0 \\ . \\ . \\ 0 \\ 1 \\ 0 \\ . \\ . \\ . \\ 0 \end{bmatrix} \leftarrow i\text{th place}$$

From this, we have,

(7A.22a) $< x_i | x_j > = \delta_{ij}$

(7A.22b) $\Sigma | x_i >< x_i | = I$

(7A.22c) $| f_n(x) > = \Sigma f_n(x_i) | x_i >$, where i = 1....n

(7A.23a) All we have to do to go from the discrete to the continuous spectrum is,

$n \rightarrow \infty$, and $\Sigma \rightarrow \int$. For instance,

$\Sigma | x_i >< x_i | = I \rightarrow \int | x >< x | dx = I$

(7A.23b) $< x_i | x_j > = \delta(x_i - x_j)$, where δ is now the Dirac Delta Function (appendix J, equation JB.1).

(7A.24) Note that < x | f > is just the projection of | f > along the basis | x >, which is just f(x).

< x | f > = f(x)

Likewise, < g | x > = g*(x).

(7A.25) The inner product (equation 7A.8) becomes,

< f | g > = ∫ < f | x >< x | g > dx (equation 7A.23a)

   = ∫ f*(x)g(x)dx (equation 7A.15)

7B. Particle on a Line

Consider a particle moving along a line in the x-direction. Every point on the line can be represented by a function of x, denoted by |Ψ(x)>. The observable would be the operator that locates the particle on the x-axis, say X. In this case we can write,

(7B.1) X|Ψ(x)> = x|Ψ(x)>

The operator X simply multiplies the function Ψ(x) by x. The next step is to find if there are eigenvalues, and what are they. To do that we would need to find the eigenvectors such that,

(7B.2) X | λ > = λ | λ >, where | λ > is the eigenket, and λ is the eigenvalue.

The equation we're looking for is then

(7B.3) $(x - \lambda) \mid \lambda > = 0$

This tells us that whenever $x \neq \lambda$, then the function $\mid \lambda >$ is zero. The only place where $\mid \lambda >$ is not zero is when $x = \lambda$. Let's plot what $\mid \lambda >$ looks like.

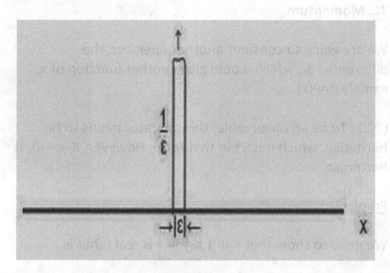

Fig. 7B.1

As the interval ε decreases to zero at $x = \lambda$, the function goes to infinity. This function is known as the *Dirac delta function*.

(7B.4) $\mid \lambda > = \delta (x - \lambda)$

Note: the area under the delta function is 1 (ε x $1/\varepsilon$).

(7B.5) To find the probability amplitude of detecting the particle at position x, we calculate the amplitude $< x \mid \Psi >$ and then square it according to the Born rule:

$P(x) = \mid < x \mid \Psi > \mid^2 = < x \mid \Psi >< \Psi \mid x > = \Psi(x)^*\Psi(x)$

183

The implication is that if Ψ(x) is any function of x, we can calculate the probability of finding a particle at x with the above mathematical structure we have constructed.

7C. Momentum

We are going to consider another operator, the differential ∂_x, which would give another function of x, namely $\partial_x\Psi(x)$.

(7C.1) To be an observable, this operator needs to be Hermitian, which it isn't in that form. However, $K = -i\partial_x$ is Hermitian.

Proof:

We need to show that $< \Psi \mid K \mid \Psi >$ is real , that is,

$< \Psi \mid K \mid \Psi > = < \Psi \mid K \mid \Psi >^*$ (equation 7A.20)

(i) By definition,

$< \Psi \mid K \mid \Psi > = \int \Psi^*(x)(-i\partial_x)\Psi(x)\,dx$
$\qquad\qquad\qquad = -i\int \Psi^*(x)\,(\partial\Psi(x)/\partial x)\,dx$

(ii) Integrating by parts (appendix E, equation EB.4), we get

$< \Psi \mid K \mid \Psi > = i\int \Psi(x)(\partial\Psi^*(x)/\partial x)\,dx$

(iii) Complex conjugate the above,

$< \Psi \mid K \mid \Psi >^* = -i\int \Psi^*(x)(\partial\Psi(x)/\partial x)\,dx$

Which the same as (i), QED.

Our next step is to find the eigenvectors of this operator K, that is,

(7C.2) $- i\partial_x \Psi(x) = k \Psi(x)$, where k is a real number

Aside from an arbitrary constant, which we can neglect for our purposes, a solution to that equation is,

(7C.3) $\Psi(x) = e^{ikx}$

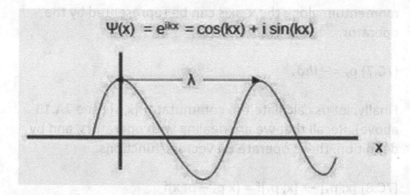

$$\Psi(x) = e^{ikx} = \cos(kx) + i \sin(kx)$$

Fig. 7C.1

These are the eigenvectors of the operator K , which are exponential functions. While the position eigenfunctions are peaks, the eigenfunctions of K extend over all spaces, oscillating everywhere with equal probability. So what is this operator? We can see from fig 7C.1 that after the wavelength λ, the wavefunction will repeat itself. For each cycle,

(7C.4) $k\lambda = 2\pi$, or $\lambda = 2\pi/k$

(7C.5) Here we appeal to history. De Broglie had hypothesized in the wave/particle duality that every wavelength was inversely proportional to the momentum, $p = h/\lambda$, where h is the Planck constant.

Substituting for λ,

(7C.6) $p = (h/2\pi) k = \hbar k$ where $\hbar = h/2\pi$

We can see that our operator K differs from the momentum operator by a factor $\hbar$. One of the most remarkable result of quantum mechanics is that momentum along the x-axis can be represented by the operator,

(7C.7) $p_x = -i\hbar\partial_x$

Finally, let us calculate the commutator [x,p] (see 7A.13 above). Recall that we are dealing with operators, and by definition, these operate on vectors/functions.

(7C.8) $[x, p_x] \rightarrow [x, p_x]f = (x\, p_x - p_x\, x)f$

$$= x(-i\hbar\partial_x)f - (-i\hbar\partial_x)(xf)$$

$$= x(-i\hbar\partial_x f) + x(i\hbar\partial_x f) + i\hbar f = i\hbar f$$

Therefore, $[x, p_x] = i\hbar$

These two operators don't commute. We see how the Heisenberg Uncertainty Principle is revealed. If we know the position of the particle with definite precision (the Dirac delta functions), then the eigenvectors of the momentum (the exponential functions) tells us we can't define its momentum as these wavefunctions are spread out equally all over the space (the x-axis). In this case, when the two operators don't commute, $[x,p] = i\hbar \neq 0$, technically, if one is in an eigenstate, the other cannot be, we say that the position and momentum of a particle are *incompatible observables*.

(7C.9) The time evolution of the wavefunction is governed by the Schrödinger equation:

$$i\hbar \frac{\partial}{\partial t}|\psi\rangle = H|\psi\rangle$$

Classically the Hamiltonian H can be written as the sum of the kinetic energy and the potential energy:

(7C.10) H = T + V (equation 2A.2)

The kinetic energy is:

(7C.11) $T = \frac{1}{2} m v^2$

It can be expressed in terms of the momentum:

(7.C12) $T = \frac{1}{2} \left(\frac{m^2}{m}\right) v^2 = \frac{p^2}{2m}$

In this form we get immediately the quantum mechanical version by using equation 7C.7. We get,

(7C.13) $T = -\frac{\hbar^2}{2m} \partial_x^2$

The Schrödinger equation in 1-D now reads as:

(7C.14)

$$i\hbar \frac{\partial}{\partial t}|\Psi(x,t)\rangle = -\frac{\hbar^2}{2m} \partial_x^2 |\Psi(x,t)\rangle + V(x,t)|\Psi(x,t)\rangle$$

We include the general case that the wavefunction could be time-dependent.

As we have mentioned at the beginning of this chapter, the mathematical formalism of QM is quite different than classical physics. We have Hermitian operators, representing observables, operating on the wave function, which contains the information of the system, such as position, momentum, etc. and with this, we can calculate probabilities according to the Born rule.

7D. The Harmonic oscillator

We shall repeat the classical case. To begin with Hooke's Law for a body oscillating under a restoring force:

(7D.1) $F = -kx$, where k is a constant for the restoring force

(7D.2) From Newton's law of motion, we have

$mx'' + kx = 0$, where x'' is the acceleration (second derivative wrt time).

We write this as,

(7D.3) $x'' + \omega^2 x = 0$,

where $\omega = (k/m)^{\frac{1}{2}}$, is the classical frequency of oscillation.

This equation is known as the wave equation. Its general solutions is,

(7D.4) $x = A\cos(\omega t) + B\sin(\omega t)$, where A and B are constants whose value will be determined by initial conditions.

We can also write the general solution, but this time slightly different,

(7D.5) $x = x_0\cos(\omega t + \phi)$, where x_0 is the amplitude and ϕ is the phase.

In the Hamiltonian formalism, we have (equation 2A.2)

(7D.6) $H = T + V$,

Again restating that H is the energy of the system, T is the kinetic energy, and V is the potential energy.

From equation 1F.24, the potential energy is,

(7D.7) $V = \frac{1}{2}m\,\omega^2\,x^2$

The kinetic energy is,

(7D.8) $T = \frac{1}{2}m\,v^2 = \frac{1}{2}m\,\dot{x}^2$

So in this case for the harmonic oscillator, equation 7D.6 becomes

(7D.9) $H = \frac{1}{2}m\dot{x}^2 + \frac{1}{2}m\omega^2 x^2$

Using $H \equiv E$, and substituting (7D.5) into (7D.9)),

(7D.10) $E = \frac{1}{2}mx_0^2\omega^2\sin^2(\omega t+\phi)+ \frac{1}{2}m\omega^2x_0^2\cos^2(\omega t+\phi)$

$\qquad = \frac{1}{2}m\,x_0^2\omega^2[\sin^2(\omega t+\phi)+\cos^2(\omega t+\phi)]$

$$= \tfrac{1}{2}m\omega^2 x_0^2$$

We see that the energy depends on the amplitude x_0 and the frequency ω, and it is independent of the phase ϕ. Since the amplitude is a continuous variable, so is the energy E.

Solving for the velocity, $\dot{x}$ in equation (7D.9),

(7D.11) $(\dot{x})^2 = 2E/m - \omega^2 x^2$

(7D.12) Or, $\dot{x} = (2E/m - \omega^2 x^2)^{\frac{1}{2}}$

Substituting (7D.10),

(7D.13) $\dot{x} = \omega (x_0^2 - x^2)^{\frac{1}{2}}$

We see that the particle will oscillate and $x = \pm x_0$ are the turning points, after which the particle speeds up until it reaches the origin, slows down until it reaches the next turning point, then reverses direction to repeat the motion in endless oscillations.

To quantize the oscillator, we must express our system into some eigenvalue equation. We could choose position, momentum or energy as a basis. For the oscillator, it is customary to choose the energy as our basis. We write,

(7D.14) $H|E> = E|E>$, where $|E>$ is the ket basis, and E is the eigenvalue, a real number. Our Hamiltonian operator H can be expressed in terms of the position operator X and the momentum operator P from equation (7D.9) as,

(7D.15) $H = P^2/2m + \frac{1}{2}m\omega^2X^2$

With equation 7C.7, repeated below,

(7D.16) $P = -i\hbar\partial_x$

And equation 7C.8, again repeated below,

(7D.17)) $[X,P] = i\hbar$

We now define the annihilation operator a and its adjoint, the creation operator $a^\dagger$, as,

(7D.18) $a = fX + igP$

(7D.19) $a^\dagger = fX - igP$

Where

(7D.20) $f = (m\omega/2\hbar)^{\frac{1}{2}}$ and $g = (2m\omega\hbar)^{-\frac{1}{2}}$. Note that f and g are not operators but rather constant, and $fg = (2\hbar)^{-1}$.

We now calculate the commutation between these two operators,

(7D.21) $[a,a^\dagger] = aa^\dagger - a^\dagger a$

$$= (fX + igP)(fX - igP) - (fX - igP)(fX + igP)$$

$$= -2i(fg)[X,P]$$

$$= 1, \text{ using (7D.17) and (7D.20)}$$

Next we calculate the operator $a^\dagger a$,

(7D.22) $a^\dagger a = (fX - igP)(fX + igP)$

$= f^2X^2 + g^2P^2 + ifg[X,P]$,

$= (m\omega/2\hbar)X^2 + (2m\omega\hbar)^{-1}P^2 + i(2\hbar)^{-1}(i\hbar)$ (using 7D.20)

$= (\hbar\omega)^{-1}(\tfrac{1}{2}m\omega^2X^2 + P^2/2m) - \tfrac{1}{2}$

$= H/\hbar\omega - \tfrac{1}{2}$, using (7D.15)

Rearranging the above,

(7D.23) $H = \hbar\omega(a^\dagger a + \tfrac{1}{2})$

$= a^\dagger a + \tfrac{1}{2}$, (from now on, $\hbar\omega = 1$)

The commutation between a and H is,

(7D.24) $[a,H] = [a, a^\dagger a + \tfrac{1}{2}]$

$= [a, a^\dagger a]$

$= aa^\dagger a - a^\dagger aa$

$= aa^\dagger a - (aa^\dagger - 1)a$ (using 7D.21)

$= a$

Similarly, the commutation between $a^\dagger$ and H is,

(7D.25) $[a^\dagger,H] = [a^\dagger, a^\dagger a + \tfrac{1}{2}]$

$$= [a^\dagger, a^\dagger a]$$

$$= a^\dagger a^\dagger a - a^\dagger a a^\dagger$$

$$= a^\dagger(aa^\dagger - 1) - a^\dagger aa^\dagger \text{ (using 7D.21)}$$

$$= -a^\dagger$$

We now get an interesting unexpected result if we consider,

$$(7D.26) \quad Ha^\dagger|E> = (a^\dagger H - [a^\dagger, H])|E>$$

$$= (a^\dagger H + a^\dagger)|E>, \text{ using (7D.25)}$$

$$= (E + 1)a^\dagger|E>, \text{ using (7D.14)}$$

Similarly, if we consider this expression,

$$(7D.27) \quad Ha|E> = (aH - [a, H])|E>$$

$$= (aH - a)|E>, \text{ using (7D.24)}$$

$$= (E - 1)a|E>, \text{ using (7D.14)}$$

We see that $a^\dagger|E>$ is an eigenstate of the Hamiltonian H, with eigenvalue E+1; and $a|E>$ is also an eigenstate of the Hamiltonian H, with eigenvalue E− 1 . By repeatingly acting the operator a or $a^\dagger$, we can get all the eigenvalues as,

$$(7D.28) \quad E + 1, E + 2, E + 3... E + \infty , \text{ (upward chain)}$$

$$(7D.29) \quad E - 1, E - 2, E - 3... E - \infty, \text{ (downward chain)}$$

IMPORTANT: The downward chain must break at some point. There must be a state |0 > that cannot be lowered any more. That is,

(7D.30) a|0> = 0.

This is how we come to the concept of a ground state or vacuum energy state, which is one of the most important concepts in QFT. And it will turn out that such a definition will cause us trouble in many situations.

Using the same scheme, we can define the concept of a particle:

(7D.31) $a^{\dagger}$|0> = N|1_k >,

where N is a normalizing factor, and |1_k > is a state containing one particle with momentum k.

For our purpose we can drop the index k. It is also why the operator $a^{\dagger}$ is called a creation operator as it creates a particle from the vacuum state. Similarly, the operator a is the annihilation operator because it annihilates a particle when acting, and if that state is the vacuum, we get equation (7D.30).

Applying the Hamiltonian on the vacuum state, (and restoring $\hbar\omega$)

(7D.32) H|0> = $\hbar\omega(a^{\dagger}a + \frac{1}{2})$|0>, using equation (7D.23)

$= \frac{1}{2} \hbar\omega$|0>, using equation (7D.30)

We can see that the vacuum energy is not zero, which will be another potential problem in QFT, where we must sum up over all modes. However this is to be expected since there are an infinite number of points in space. But we will see later on that restraining our arguments to energy density, rather than energy, we can avoid that problem.

For all the excited states,

(7D.33) $|n> = (a^{\dagger})^n|0>$.

7E. Angular Momentum

Recall the definition of angular momentum as,

(7E.1) $\mathbf{L} = \mathbf{r} \times \mathbf{p}$ (equation 1D.4b)

Where in Cartesian coordinates, $\mathbf{L} = (L_x, L_y, L_z)$, $\mathbf{r} = (x, y, z)$, $\mathbf{p} = (p_x, p_y, p_z)$,

This gives the following components of the quantized angular momentum (using equation 7C.7):

(7E.2a) $L_x = yp_z - zp_y = = -i\hbar(y\frac{\partial}{\partial z} - z\frac{\partial}{\partial y})$

(7E.2b) $L_y = zp_x - xp_z = = -i\hbar(z\frac{\partial}{\partial x} - x\frac{\partial}{\partial z})$

(7E.2c) $L_z = xp_y - yp_x = = -i\hbar(x\frac{\partial}{\partial y} - y\frac{\partial}{\partial x})$

Before proceeding, let's establish the following identity,

(7E.3) For any arbitrary three operators A, B, C:

[AB,C] = A[B,C] + [A,C]B

Proof:

(i) Expand the LHS $=$ ABC $-$ CAB

Add and subtract ACB,

(ii) LHS $=$ ABC $-$ CAB $+$ ACB $-$ ACB

Combine 1st and 4th term, then 2nd and 3rd terms:

(iii) LHS $=$ ABC $-$ ACB $+$ ACB $-$ CAB
$=$ A (BC $-$ CB) $+$ (AC $-$ CA)B
$=$ A[B,C] $+$ [A,C]B   QED

An important result is,

(7E.4) $\left[L_i , L_j \right] = i\hbar \sum_{k=1}^{k=3} \varepsilon_{ijk} L_k$

Where i, j run from 1 to 3, and L_1, L_2, L_3 stand for L_x, L_y, L_z respectively. And ε_{ijk} is an anti-symmetric tensor of rank 3 with the following properties:

(7E.4a) $\varepsilon_{123} = 1$

(7E.4b) ε_{ijk} changes sign whenever two indices are exchanged. Therefore, no two indices can be equal.

To prove 7E.4, we start with [L_x, L_y] and equations 7E.2,

(i) $\left[L_x, L_y \right] = $ [$yp_z - zp_y$, $zp_x - xp_z$]

$$= [\,yp_z\,,zp_x\,] - [\,yp_z\,,xp_z\,]$$
$$-[\,zp_y\,,zp_x\,] + [zp_y\,,xp_z\,]$$

Recall that the only operators that fail to commute are x with p_x, y with p_y and z with p_z. Therefore both 2nd and 3rd terms vanish. We are left with,

(ii) $[L_x, L_y] = [\,yp_z\,,zp_x\,] + [zp_y\,,xp_z\,]$

In the 1st bracket, we can take out y and p_x as both commute with z and p_z. Similarly for the 2nd bracket, x and p_y both commute with z and p_z. We then have,

(iii) $[L_x, L_y] = yp_x[\,p_z,z] + xp_y\,[z, p_z\,]$

Using equation 7C.8, we then get,

(iv) $[L_x, L_y] = yp_x(-i\hbar) + xp_y(i\hbar)$
$$= (i\hbar)(\,xp_y - yp_x)$$

With equation 7E.2c,

(v) $[L_x, L_y] = i\hbar\, L_z$

(vi) With cyclic permutations of the indices (x → y, y → z, z → x), we get equation 7E.4

Other important commutation relationships are,

(7E.5) $[\,L^2\,, L_i\,] = [\,L_i\,, p^2\,] = [\,L_i\,, r^2\,] = [\,L_i\,, \mathbf{r} \cdot \mathbf{p}\,] = 0$

Where again $i = 1,2,3$ and $L^2 = L_x^2 + L_y^2 + L_z^2$
We define the raising and lowering operators as,

(7E.6) $L_+ \equiv L_x + iL_y$ and $L_- \equiv L_x - iL_y$

Note that that $L_\mp$ is the hermitian conjugate of $L_\pm$, that is, $(L_\mp)^* = L_\pm$.

The commutator of L_z with L_+ is,

(7E.7) $[L_z, L_+] = [L_z, L_x + iL_y] = [L_z, L_x] + i[L_z, L_y]$

Using equation 7E.4,

$[L_z, L_+] = i\hbar L_y + i(-i\hbar L_x) = \hbar(L_x + iL_y) = \hbar L_+$

Similarly,

(7E.8) $[L_z, L_-] = -\hbar L_-$

Using the first equation in 7E.5, it can be easily shown that,

(7E.9) $[L^2, L_+] = [L^2, L_-] = 0$

(7.E10) We can now make the claim that if a function f is an eigenfunction of L^2 ($L^2 f = \lambda f$, by definition) and L_z ($L_z f = \mu f$, by definition), where λ and μ are constants then it is also an eigenfunction of $L_\pm f$.

Proof: Consider

(a) $L^2(L_+ f) = L_+ L^2 f$, (using equation 7E.9)
$= L_+ \lambda f$, (def. of eigenfunction)
$= \lambda (L_+ f)$

Consider:

(b) $L_z (L_+f) = \{-[L_z, L_+] + \hbar L_+\}f + L_z(L_+f),$

$$= \{-L_zL_+ + L_+L_z + \hbar L_+\}f + L_z(L_+f)$$
$$= L_+L_zf + \hbar L_+f, \text{ (first and last terms cancel)}$$
$$= L_+\mu f + \hbar L_+f, \text{ (def. of eigenfunction)}$$
$$= (\mu + \hbar)(L_+f)$$

(c) Similarly, $L^2(L_-f) = \lambda(L_+ f)$
And $\quad\quad L_z (L_-f) = (\mu - \hbar)(L_-f)$

We can see that $(L_\pm f)$ is an eigenfunction of L_z but with new eingenvalues $\mu \pm \hbar$.

We call L_+ the raising ladder operator because it raises the eigenvalue of L_z by $\hbar$; likewise, L_- is the lowering operator because it decreases the eigenvalue of L_z by $\hbar$.

So what we have is a ladder of states, each rung being separated by one unit of $\hbar$ in the eigenvalue of L_z. We apply the raising operator L_+ to ascend the ladder and lowering operator L_- to descend. But this process cannot be applied infinitely. At one point the z-component of the vector **L** will exceed the total length of vector **L**!? That is unrealistic. So there must exist a top rung such that,

(7E.11) $L_+f_{top} = 0$

Let $\hbar\ell$ be the eigenvalue of L_z at this top rung. That is,

(7E.12) $L_zf_{top} = (\hbar\ell) f_{top}$ and $L^2f_{top} = \lambda f_{top}$

Consider:

$$(7E.13) \quad L_-L_+ = (L_x - iL_y)(L_x + iL_y)$$
$$= L_x^2 + L_y^2 + i\,(L_xL_y - L_yL_x)$$
$$= L_x^2 + L_y^2 + i\,(i\hbar L_z), \text{ using 7E.4}$$
$$= L^2 - L_z^2 - \hbar L_z, \text{ definition of } L^2 \text{ in 7E.5}$$

Similarly,

$$(7E.14) \quad L_+L_- = L^2 - L_z^2 + \hbar L_z,$$

We write equation 7E.13-14 as,

$$(7E.15a) \quad L^2 = L_-L_+ + L_z^2 + \hbar L_z$$

$$(7E.15b) \quad L^2 = L_+L_- + L_z^2 - \hbar L_z$$

Now we apply 7E.15a on f_{top}. We get,

$$(7E.16) \quad L^2 f_{top} = (L_-L_+ + L_z^2 + \hbar L_z)f_{top}$$
$$\lambda f_{top} = (0 + (\hbar\ell)^2 + \hbar(\hbar\ell)\,f_{top}, \text{ using 7E.11-12}$$

Therefore,

$$(7E.17) \quad \lambda = \hbar^2 \ell\,(\ell + 1)$$

From the same arguments we have a bottom rung such that,

$$(7E.18a) \quad L_- f_{bottom} = 0$$

Let $\hbar\ell'$ be the eigenvalue of L_z at this bottom rung. That is,

(7E.18b) $L_z f_{bottom} = (\hbar \ell') f_{bottom}$; $L^2 f_{bottom} = \lambda f_{bottom}$

Applying 7E.15b on f_{bottom}, we now get,

(7E.19) $L^2 f_{bottom} = (L_+ L_- + L_z^2 - \hbar L_z) f_{bottom}$
$\lambda f_{bottom} = (0 + (\hbar \ell')^2 - \hbar (\hbar \ell')) f_{bottom}$,

Therefore,

(7E.20) $\lambda = \hbar^2 \ell'(\ell' - 1)$

Equating equations 7E.17 and 7E.20, we have,

(7E.21) $\ell (\ell + 1) = \ell'(\ell' - 1)$

Either we have $\ell' = \ell + 1$, which is not possible as the bottom rung would be higher than the top rung!! Or we have,

(7E.22) $\ell' = - \ell$.

We label the eigenvalues of L_z as $m\hbar$, and we see that in terms of $\hbar$, the eigenvalues of L_z will go from $- \ell$ to ℓ, in steps of N, where N is an integer (0,1,2...).

(7E.23) $m = - \ell, - \ell+1, ... , \ell-1, \ell$

We now have the values of m in terms of the values of ℓ, but what about the values of ℓ itself? We know that m increases in N steps. From equation 7E.22 we can calculate the difference between the top and bottom rungs:

(7E.24) $\ell - \ell' = \ell - (-\ell) = 2\ell = N$, or $\ell = \dfrac{N}{2}$

In terms of $\hbar$, $\ell = 0, \dfrac{1}{2}, 1, \dfrac{3}{2}, ...$ That is, ℓ is either a half-integer or an integer. We can now label our eigenfunctions in 7E.10 as $|\ell\, m >$. To resume what we have so far;

(7E.25) $L^2 |\ell\, m > = \hbar^2 \ell\, (\ell + 1)$

(7E.26) $L_z |\ell\, m > = \hbar m |\ell\, m >$

Note: The raising and lowering operators change the value of m by one unit.

Claim: $L_\pm |\ell\, m > = C_\pm |\ell\, (m\pm1) >$, where $C_\pm$ is some constant to be determined later.

Proof: Consider equation 7E.7, rewritten below,

(i) $[L_z, L_+] = \hbar L_+$

Apply this to the eigenfunctions $|\ell\, m >$. We have,

(ii) $\hbar L_+ |\ell\, m > = [L_z, L_+] |\ell\, m >$
$= (L_z L_+ - L_+ L_z) |\ell\, m >$
$= L_z L_+ |\ell\, m > - L_+ \hbar m |\ell\, m >$, using 7E.26

Rearranging,

(iii) $L_z L_+ |\ell\, m > = \hbar L_+ |\ell\, m > + L_+ \hbar m |\ell\, m >$
$= \hbar (m+1) L_+ |\ell\, m >$

So we can see that L_z acting on eigenfunction $L_+ |\ell\, m >$ is proportional to the eigenfunction $|\ell\, m +1 >$. Similarly, we would get that L_z acting on the function $L_- | \ell\, m >$ is proportional to the function $| \ell\, m - 1 >$. Therefore,

(iv) $L_\pm | \ell\, m > = C_\pm | \ell\, (m \pm 1) >$, QED for the first part.

To determine $C_\pm$, note that $(L_\mp)^* = L_\pm$, (see equation 7E.6).

(v) From $L_+ | \ell\, m > = C_+ | \ell\, (m + 1) >$, we take the conjugate and get,

(vi) $< \ell\, m | L_- = C_+^* < \ell\, (m + 1) |$

Multiply equations (v) and (vi). The RHS gives,

(vii) RHS $= C_+ C_+^* < \ell\, (m + 1) | \ell\, (m + 1) > = C_+^2$

For the LHS, we use equation 7E.13,

(viii) LHS $= < \ell\, m | L_- L_+ | \ell\, m >$
$= < \ell\, m | L^2 - L_z^2 - \hbar L_z | \ell\, m >$
$= \hbar^2 \ell\, (\ell + 1) - \hbar^2 m^2 - \hbar^2 m$
$= \hbar^2 [\ \ell^2 + \ell - m^2 - m]$
$= \hbar^2 [\ell\, (\ell + 1) - m(m + 1)]$

Equating (vii) and (viii), and taking the square root,

(ix) $C_+ = \hbar\, [\ell\, (\ell + 1) - m(m + 1)]^{1/2}$

Similarly, we get C_- by substituting $m \to -m$

(x) $C_- = \hbar\, [\ell\, (\ell + 1) - m(m - 1)]^{1/2}$

Therefore,

(7E.27) $L_\pm |\ell\, m > = \hbar\, [\ell\,(\ell + 1) - m(m \pm 1)]^{\frac{1}{2}} |\ell\,(m \pm 1) >$

7F. Spin

The algebra of spin is an exact copy of orbital angular momentum. So we carry its algebraic structure. We make a brief change of notation: $L \to S$; $\ell \to s$. From equations 7E.25-26,

(7F.1) $S^2 |s\, m > = \hbar^2 s(s + 1) |s\, m >$

(7F.2) $S_z |s\, m > = \hbar m |s\, m >$

Similarly the ladder operators are:

(7F.3) $S_+ \equiv S_x + iS_y$ and $S_- \equiv S_x - iS_y$

With similar results (equations 7E.15a -b),

(7F.4) $S^2 = S_\pm\, S_\mp + S_z^2 \mp \hbar S_z$

Applying these to the quantum states (equation 7E.27), we have

(7F.5) $S_\pm |s\, m >$
$$= \hbar \sqrt{s(s + 1) - m(m \pm 1)} \; |s\,(m \pm 1) >$$

Now we will look at the case s = ½, for the major reason that most of the matter in the universe is made up of electrons, protons and neutrons, including quarks and leptons, all having spin ½. For s =½, m can only take the values of +½ and - ½ (see 7E.23). So we have only two eigenfunctions. In the literature you'll often see them as $|½\ ½ >$ and $|½\ -½ >$. Since we will not be concerned with the total spin for a little while, these are going to be denoted as $|\uparrow >$ or spin up, and $|\downarrow >$ or spin down, respectively. The general state of a spin-½ particle can then be expressed as a two-element column matrix – also called a spinor:

$$(7F.6)\quad \chi = \begin{pmatrix} a \\ b \end{pmatrix} = a \begin{pmatrix} 1 \\ 0 \end{pmatrix} + b \begin{pmatrix} 0 \\ 1 \end{pmatrix}$$

Define:

$$(7F.7)\quad \chi_+ = \begin{pmatrix} 1 \\ 0 \end{pmatrix} \text{ for spin up}$$

And $\quad \chi_- = \begin{pmatrix} 0 \\ 1 \end{pmatrix}$ for spin down

From equation 7F.1, we can evaluate,

$$(7F.8a)\quad S^2\chi_+ = \hbar^2 s(s + 1)\,\chi_+$$
$$= \hbar^2 ½(½ + 1)\,\chi_+$$
$$= \frac{3}{4}\hbar^2\chi_+$$

Similarly,

$$(7F.8b)\quad S^2\chi_- = \frac{3}{4}\hbar^2\chi_-$$

Now if the states are column matrices, the operators must be 2 x 2 matrices for the theory to be consistent. For S^2 we write,

(7F.9) $S^2 = \begin{pmatrix} c & d \\ e & f \end{pmatrix}$

Our next job is to determine the unknown elements. First we use, equation 7F.8a,

$$\begin{pmatrix} c & d \\ e & f \end{pmatrix}\begin{pmatrix} 1 \\ 0 \end{pmatrix} = \tfrac{3}{4}\hbar^2 \begin{pmatrix} 1 \\ 0 \end{pmatrix} \rightarrow \begin{pmatrix} c \\ e \end{pmatrix} = \begin{pmatrix} \tfrac{3}{4}\hbar^2 \\ 0 \end{pmatrix}$$

This means that $c = \tfrac{3}{4}\hbar^2$ and $e = 0$.

Similarly, using equation 7F.8b, $d = 0$ and $f = \tfrac{3}{4}\hbar^2$. And so our 2 x 2 matrix for S^2 is:

(7F.10) $S^2 = \tfrac{3}{4}\hbar^2 \begin{pmatrix} 1 & 0 \\ 0 & 1 \end{pmatrix}$

From equation 7F.2, we can determine the matrix for S_z.

(7F.11) $S_z \chi_+ = \tfrac{\hbar}{2}\begin{pmatrix} 1 \\ 0 \end{pmatrix}$; $S_z \chi_- = -\tfrac{\hbar}{2}\begin{pmatrix} 0 \\ 1 \end{pmatrix}$

From these results, we get,

(7F.12) $S_z = \tfrac{\hbar}{2} \begin{pmatrix} 1 & 0 \\ 0 & -1 \end{pmatrix}$

To determine the ladder operators, we use equation 7F.5, rewritten below, taking into account that s = ½, m = ½ for χ_+, spin up; and m= −½ for χ_-, spin down.

(7F.13a) $S_+\chi_+ = \hbar \sqrt{½(½ + 1) - ½(½ + 1)}\ \chi_- = 0$

Similarly,

(7F.13b)

$$S_-\chi_- = \hbar \sqrt{\tfrac{1}{2}(\tfrac{1}{2}+1)-(-\tfrac{1}{2})(-\tfrac{1}{2}-1)}\ \chi_+ = 0$$

Also,

(7F.13c)

$$S_+\chi_- = \hbar \sqrt{\tfrac{1}{2}(\tfrac{1}{2}+1)-(-\tfrac{1}{2})(-\tfrac{1}{2}+1)}\ \chi_+ = \hbar\chi_+$$

And

(7F.13d)

$$S_-\chi_+ = \hbar \sqrt{\tfrac{1}{2}(\tfrac{1}{2}+1)-\tfrac{1}{2}(\tfrac{1}{2}-1)}\ \chi_+ = \hbar\chi_-$$

This gives the form for the ladder operators as,

(7F.14a) $S_+ = \hbar \begin{pmatrix} 0 & 1 \\ 0 & 0 \end{pmatrix}$

(7F.14b) $S_- = \hbar \begin{pmatrix} 0 & 0 \\ 1 & 0 \end{pmatrix}$

From equation 7F.3 we have,

(7F.15a) $S_x = \tfrac{1}{2}(S_+ + S_-) = \tfrac{\hbar}{2} \begin{pmatrix} 0 & 1 \\ 1 & 0 \end{pmatrix}$

(7F.15b) $S_y = \tfrac{-i}{2}(S_+ - S_-) = \tfrac{\hbar}{2} \begin{pmatrix} 0 & -i \\ i & 0 \end{pmatrix}$

Since S_x, S_y and S_z all carry a factor $\tfrac{\hbar}{2}$ it is customary to write $S = \tfrac{\hbar}{2}\sigma$ and define the Pauli matrices as,

(7F.16) $\sigma_x \equiv \begin{pmatrix} 0 & 1 \\ 1 & 0 \end{pmatrix}, \sigma_y \equiv \begin{pmatrix} 0 & -i \\ i & 0 \end{pmatrix}, \sigma_z \equiv \begin{pmatrix} 1 & 0 \\ 0 & -1 \end{pmatrix}$

The question arises: what are the possible values of the total spin when you combine the spins of two spin ½ particles?

We have,

(7F.17) $\boldsymbol{S} = \boldsymbol{S}^{(1)} + \boldsymbol{S}^{(2)}$

We now have two spinors (equation 7F.6) denoted as χ_1 and χ_2. We also know that each particle has two components along the z-axis. That is,

$$
\begin{aligned}
\text{(7F.18) } S_z \chi_1 \chi_2 &= \left(S_z^{(1)} + S_z^{(2)} \right) \chi_1 \chi_2 \\
&= (S_z^{(1)} \chi_1) \chi_2 + \chi_1 (S_z^{(2)} \chi_2) \\
&= (\hbar m_1 \chi_1) \chi_2 + \chi_1 (\hbar m_2 \chi_2) \\
&= \hbar (m_1 + m_2) \chi_1 \chi_2
\end{aligned}
$$

We see that the eigenvalues of the z-components simply add. In the literature this is represented as:

$$
\begin{aligned}
\text{(7F.19) } & \mid \uparrow\uparrow >, m = 1 \\
& \mid \uparrow\downarrow >, m = 0 \\
& \mid \downarrow\uparrow >, m = 0 \\
& \mid \downarrow\downarrow >, m = -1
\end{aligned}
$$

There are two things noticeable: (a) the spin should increase in integral steps, which it isn't doing; and (b) there are two states with $m = 0$. Another way to look at this is with the ladder operators (equations 7F.3). For instance, applying S_- to the state $\mid \uparrow\uparrow >$,

(7F.20) $S_- \mid \uparrow\uparrow> = \left(S_-^{(1)} + S_-^{(2)} \right) \mid \uparrow\uparrow>$

$\qquad = S_-^{(1)} \mid \uparrow\uparrow> + S_-^{(2)} \mid \uparrow\uparrow>$

$\qquad = \hbar(\mid \downarrow\uparrow> + \mid \uparrow\downarrow>)$

Notice that inside the bracket, this is one of the ($m = 0$) states. Obviously, the other ($m = 0$) state would be of the type:

$$\mid \downarrow\uparrow> - \mid \uparrow\downarrow>$$

This is best represented in the $\mid s\ m >$ notation as the followings:

(7F.21a) In the case of $s = s_1 + s_2 = $ ½ + ½ = 1, the two particle's total spin are aligned, we have the triplet:

$\mid 1\ 1\ > = \ \mid \uparrow\uparrow>, m = 1$

$\mid 1\ 0\ > = \frac{1}{\sqrt{2}}(\mid \uparrow\downarrow> + \mid \downarrow\uparrow>), m = 0$

$\mid 1-1\ > = \ \mid \downarrow\downarrow>, m = -1$

(Note: $\frac{1}{\sqrt{2}}$ is a normalizing factor)

(7F.21b) In the case of the singlet, in which the two particle's total spin are anti-aligned ($s = 0$):

$\mid 0\ 0\ > = \frac{1}{\sqrt{2}}(\mid \uparrow\downarrow> - \mid \downarrow\uparrow>), m = 0$

When combined, two particles of spin ½ either aligned parallel to each other, or anti-aligned. The total spin is,

(7F.22) $s = s_1 + s_2, \ s_1 + s_2 - 1, s_1 + s_2 - 2 \dots \mid s_1 - s_2 \mid$

7G. Bosons and Fermions

Suppose we have a system of two distinguishable particles 1 and 2, at $x_1 = a$ and $x_2 = b$, respectively. We denote this system by,

(7G.1) $\,|\psi(a,b)> = \,|\,ab>$

We swap the two particles. We then have,

(7G.2) $\,|\psi(b,a)> = \,|\,ba>$

We repeat the experiment but this time with identical particles. Are these two quantum states identical? In Classical physics, we would answer by a definitive yes. But in QM, this is not that clear. At most we can say is that they are equivalent, that is,

(7G.3) $\,|\psi(a,b)> \rightarrow \alpha\,|\psi(b,a)>$

Where α is a complex number. If we swap again, we get the following:

(7G.4) $\,|\psi(a,b)> \rightarrow \alpha\,|\psi(b,a)> \rightarrow \alpha^2\,|\psi(a,b)>$

This means that

(7G.5) $\alpha = \pm 1$

We can construct two fundamental quantum states. Those states that are symmetric:

(7G.6a) $\,|\psi(a,b)>_s = \,|\,ab> + \,|\,ba>$

And those that are anti-symmetric:

$$(7G.6b) \quad |\psi(a,b)>_A = |ab> - |ba>$$

It turns out that in nature particles are either bosons, which have quantum states that are symmetric; and fermions, which are anti-symmetric.

We can now deduce a fundamental property of the fermions. Consider a system of two fermions f_1 and f_2:

$$(7G.7) \quad |\psi(f_1,f_2)>_A = |f_1f_2> - |f_2f_1>$$

Suppose it's the same species of fermions: $f_1 = f_2 = f$. Then we get,

$$(7G.8) \quad |\psi(f,f)>_A = |ff> - |ff> = 0$$

This is the well-known Pauli Exclusion Principle: no two identical fermions can occupy the same quantum state. We will say more later on about bosons and the Klein-Gordon equation, and about fermions and the Dirac equation (Chapter 10).

Chapter 8

Quantum Statistical Mechanics

Before arriving at the different statistics mentioned in the previous section, we must develop the following situational concepts: stationary states; a particle in an infinite square well; followed by a particle in a rectangular solid; and lastly, the general case of a particle occupying n quantum states with d_n degeneracies in a box.

8A. Stationary States

Recall the Schrödinger equation (7C.14) in one-dimension:

$$(8A.1) \quad i\hbar \frac{\partial}{\partial t} \Psi(x,t) = -\frac{\hbar^2}{2m} \partial_x^2 \Psi(x,t) + V(x)\, \Psi(x,t)$$

What we want is a wave function that is no longer time-dependent. We can solve this type of equation using the mathematical method of separation of variables. That is, we look for solution of the type,

$$(8A.2)\ \Psi(x,t) = \psi(x)\, \varphi(t)$$

Where the function Ψ is separable into a function purely x-dependent, $\psi(x)$, and purely t-dependent, $\varphi(t)$. We take the derivative with respect to time once, then twice with respect to x:

$$(8A.3)\ \frac{\partial \Psi}{\partial t} = \psi \frac{d\varphi}{dt} \ ; \ \frac{\partial^2 \Psi}{d^2 x} = \varphi \frac{d^2 \psi}{d^2 x}$$

We substitute these results into equation 8A.1:

(8A.4) $i\hbar\psi\dfrac{d\varphi}{dt} = -\dfrac{\hbar^2}{2m}\varphi\dfrac{d^2\psi}{d^2x} + V\psi\varphi$

We now divide both sides by $\psi\varphi$:

(8A.5) $i\hbar\dfrac{1}{\varphi}\dfrac{d\varphi}{dt} = -\dfrac{\hbar^2}{2m}\dfrac{1}{\psi}\dfrac{d^2\psi}{d^2x} + V$

This is what we were aiming for: the LHS is entirely t-dependent; while the RHS is entirely x-dependent. This can only be true if both sides are equal to a constant. We shall name that constant E for reasons that will become apparent. So we have,

(8A.6a) $i\hbar\dfrac{1}{\varphi}\dfrac{d\varphi}{dt} = E \quad \rightarrow \quad i\hbar\dfrac{d\varphi}{dt} = E\varphi$

(8A.6b) $-\dfrac{\hbar^2}{2m}\dfrac{1}{\psi}\dfrac{d^2\psi}{d^2x} + V = E$

$\rightarrow -\dfrac{\hbar^2}{2m}\dfrac{d^2\psi}{d^2x} + V\psi = E\psi$

From the first equation we get an immediate solution,

(8A.7) $\varphi(t) = e^{-iEt/\hbar}$

The second equation, also called the time-independent Schrödinger equation (TISE), can only be solved when the potential V(x) is specified. We will look at a specific case in the next section.

Some general comments:

(i) Regardless that we are dealing with stationary states, the wave function is still time-dependent,

214

$$\Psi(x, t) = \psi(x)\, \varphi(t) = \psi(x)e^{-iEt/\hbar}$$

However the probability density is not time dependent (equation 7B.5):

$$|< \Psi(x, t) \mid \Psi(x, t) >|^2 = \Psi^*\Psi = \psi^* e^{+iEt/\hbar}\, \psi e^{-iEt/\hbar}$$

$$= \mid \psi(x) \mid^2$$

(ii) The time-independent Schrödinger equation (TISE) can be specified as,

$$H\psi = E\psi$$

Where the Hamiltonian $H = -\dfrac{\hbar^2}{2m}\dfrac{d^2\psi}{d^2x} + V$

(iii) The most general solution is a linear combination taking the form,

$$\Psi(x, t) = \sum_{n=1}^{\infty} c_n \psi_n(x)e^{-iE_n t/\hbar}$$

Thus there is a different wave function for each allowed energy,

$\psi_1 \rightarrow E_1\,, \psi_2 \rightarrow E_2\,, \psi_3 \rightarrow E_3$ and so on ...

8B. The Infinite Square Well

Suppose we have a specific potential such as,

(8B.1) V (x) = 0, if $0 \leq x \leq a$
$\qquad\quad = \infty$, otherwise

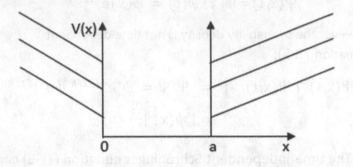

A particle is free except at the two ends where an infinite force prevents it from escape.

Inside the well, the time-independent Schrödinger equation is (from equation 8A.6b, with V(x) = 0)

(8B.2a) $-\frac{\hbar^2}{2m}\frac{d^2\psi}{d^2x} = E\psi$

$$\rightarrow \frac{d^2\psi}{d^2x} = -k^2\,\psi\,,$$

(8B.2b) Where $k \equiv \frac{\sqrt{2mE}}{\hbar}$

The general solution is,

(8B.3) $\psi(x) = A\,sinkx + Bcoskx$

Where A and B are constants to be determined by boundary conditions. From Fig.8B.1, we require that $\psi(0) = \psi(a) = 0$. At positon 0 we have,

(8B.4) $\psi(0) = A\,sin0 + Bcos0 = B = 0$

We are left with,

$$\rightarrow \psi = A \sin kx$$

At position a, we get

(8B.5) $\psi(a) = 0 = A \sin ka$

$$\rightarrow ka = 0, \pm\pi, \pm2\pi, \pm3\pi...$$

The solution k=0 can be discarded as it would imply the trivial solution $\psi(x) = 0$. The negative sign can be absorbed into the constant A. The distinct solutions are then,

(8B.6) $k_n = \frac{n\pi}{a}$ with $n = 1,2,3 ...$

The allowed energies are then quantized,

(8B.7) $E_n = \frac{\hbar^2 k_n^2}{2m} = \frac{\hbar^2 \pi^2}{2ma^2} n^2$ (using equation 8.2b)

We normalize the wave function by imposing the condition that,

(8B.8) $\int_0^a |\psi(x)|^2 dx = 1$

$$\rightarrow \int_0^a |A|^2 \sin^2(kx) \, dx = |A|^2 \left(\frac{x}{2} - \frac{\sin 2kx}{4k}\right) \Big|_0^a = |A|^2 \frac{a}{2}$$

(Using appendix C, equation CC.7b)

$$\rightarrow |A|^2 = \frac{2}{a}$$

We then have the solutions inside the well:

(8B.9) $\psi_n(x) = \sqrt{\frac{2}{a}} \sin(\frac{n\pi}{a}x)$

Some general comments:

(i) The lowest energy level, n=1, is called the ground state. The others are referred as the excited states.
(ii) With respect to the center of the well, the wave function alternates: ψ_1 is even, ψ_2 is odd, ψ_3 is even, and so on...

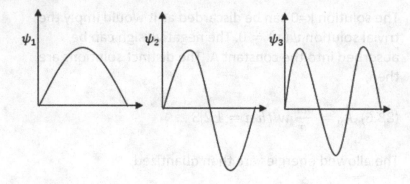

Fig. 8B.2

(iii) They are orthogonal and normalized ≡ orthonormal,

$$\int \psi_m^*(x)\psi_n(x)dx = \delta_{mn}$$

(iv) They forms a complete set, that is, any arbitrary function f(x) can be expressed as a linear combination of them:

$$f(x) = \sum_{n=1}^{\infty} c_n \psi_n$$

Where the c's can be easily evaluated as,

$$\int \psi_m^*(x) f(x) dx = \sum_{n=1}^{\infty} c_n \int \psi_m^*(x) \psi_n(x) dx$$

$$= \sum_{n=1}^{\infty} c_n \, \delta_{mn}$$

$$= c_m$$

The most general solution to the Schrödinger equation for an infinite well is,

(8B.10)

$$\Psi(x,t) = \sum_{n=1}^{\infty} c_n \sqrt{\frac{2}{a}} \sin(\frac{n\pi}{a}x) e^{-i\frac{n^2\hbar^2\pi}{2ma^2}t}$$

8C. Particle in a Rectangular Solid

What we have done so far with the infinite well is a preliminary study for a particle in a 3-D solid box. So now we have,

$$\psi(x) \rightarrow \psi(x,y,z)$$

$$V(x) \rightarrow V(x,y,z)$$

Equation 8B.1 now reads,

(8C.1) $V(x,y,z) = 0$, if $0 \le x \le l_x$, $0 \le y \le l_y$, $0 \le z \le l_z$
$\qquad = \infty$, otherwise

219

The separability of the wave function can be written in Cartesian coordinates as

(8C.2) $\psi(x, y, z) = X(x)Y(y)Z(z)$

Inside the box where the potential energy is zero, the time-independent Schrödinger equation 8A.6b now splits into three parts:

(8C.3a) $- \dfrac{\hbar^2}{2m} \dfrac{d^2 X}{d^2 x} = E_x X$

(8C.3b) $- \dfrac{\hbar^2}{2m} \dfrac{d^2 Y}{d^2 x} = E_y Y$

(8C.3c) $- \dfrac{\hbar^2}{2m} \dfrac{d^2 Z}{d^2 x} = E_z Z$

Where $E = E_x + E_y + E_z$

Similarly we define:

(8C.4a) $k_x \equiv \dfrac{\sqrt{2mE_x}}{\hbar}$

(8C.4b) $k_y \equiv \dfrac{\sqrt{2mE_y}}{\hbar}$

(8C.4c) $k_z \equiv \dfrac{\sqrt{2mE_z}}{\hbar}$

The general solution will take the general form,

(8C.5a) $X(x) = A_x \sin k_x x + B_x \cos k_x x$
(8C.5b) $Y(y) = A_y \sin k_y y + B_y \cos k_y y$
(8C.5c) $Z(z) = A_z \sin k_z z + B_z \cos k_z z$

The boundary conditions now extend into 3-D as,

(8C3.6a) X(0) = Y(0) = Z(0) = 0 $\rightarrow$ B_x = B_y = B_z = 0

(8C.6b) And X(l_x) = Y(l_y) = Z(l_z) = 0

$\rightarrow k_x = \frac{n_x \pi}{l_x}$ with $n_x = 1,2,3 \dots$

$\rightarrow k_y = \frac{n_y \pi}{l_y}$ with $n_y = 1,2,3 \dots$

$\rightarrow k_z = \frac{n_z \pi}{l_z}$ with $n_z = 1,2,3 \dots$

The normalized wave functions read as,

(8C.7) $\psi_{n_x n_y n_z}$

$$= \sqrt{\frac{8}{l_x l_y l_z}} \, \sin(\frac{n_x \pi}{l_x}x)\sin(\frac{n_y \pi}{l_y}y)\sin(\frac{n_z \pi}{l_z}z)$$

And the allowed quantized energies,

(8C.8) $E_{n_x n_y n_z} = \frac{\hbar^2}{2m}(\frac{n_x^2}{l_x^2} + \frac{n_y^2}{l_y^2} + \frac{n_z^2}{l_z^2}) = \frac{\hbar^2 k^2}{2m}$

Where k is the magnitude of the wave vector **k** = (k_x, k_y, k_z). From equation 8C.6b, the values of the components of **k** are as follows:

(8C.9a) $k_x = \frac{\pi}{l_x}, \frac{2\pi}{l_x}, \frac{3\pi}{l_x} \dots$

(8C.9b) $k_y = \frac{\pi}{l_y}, \frac{2\pi}{l_y}, \frac{3\pi}{l_y} \dots$

(8C.9c) $k_z = \frac{\pi}{l_z}, \frac{2\pi}{l_z}, \frac{3\pi}{l_z} \dots$

We can map this vector on a 3-D k-space, each point spaced by a unit quantity. Fig. 8C.1 depicts a rectangular block on this space, occupying one octant.

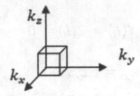

Fig. 8C.1

We can say that each quantum state occupies a volume of,

(8C.10) $V_{quantum\ state} = \dfrac{\pi}{l_x}\dfrac{\pi}{l_y}\dfrac{\pi}{l_z} = \dfrac{\pi^3}{V}$

Where $V = l_x l_y l_z$ is the volume of an object in that k-space. Consider the number of atoms N, which could be as large as Avogadro's number, each atom contributing q electrons. As fermions each electron must have a distinct quantum number. However, we can put two in a quantum state with opposite spin. We can fill up an octant of a sphere of radius k_F — we can replace the rectangular box by a sphere without any loss of generality. We have Nq electrons or $\dfrac{Nq}{2}$ pairs of electrons in a quantum state of volume $\dfrac{\pi^3}{V}$. This equal the volume of an octant of a Fermi volume:

(8C.11) $\dfrac{1}{8}\left(\dfrac{4}{3}\pi k_F^3\right) = \dfrac{Nq}{2}\dfrac{\pi^3}{V}$

Solving for k_F,

(8C.12) $k_F = (3\dfrac{Nq}{V}\pi^2)^{1/3}$
$= (3\rho\,\pi^2)^{1/3}$

Where $\rho \equiv \frac{Nq}{V}$, is the free electron density.

The corresponding Fermi energy (equation 8C.8) is,

(8C.13) $E_F = \frac{\hbar^2 k_F^2}{2m} = \frac{\hbar^2}{2m} (3\rho \, \pi^2)^{2/3}$

The total energy can be calculated as follows: Consider a shell of thickness dk (Fig. 8C.2) on one octant. It contains a volume of,

(8C.14) $V_{shell} = \frac{1}{8} (4\pi k^2) dk = \frac{1}{2} \pi k^2 dk$

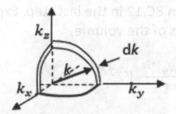

Fig. 8C.2 depicts an octant of a sphere in k-space

The number of electron states is then (two per states):

(8C.15) $N_{electron\ states} = 2 \dfrac{V_{shell}}{V_{quantum\ state}}$

$$= 2 \dfrac{\left(\frac{1}{2} \pi k^2 dk\right)}{\frac{\pi^3}{V}}$$

$$= \frac{V}{\pi^2} k^2 dk$$

For a given state (equation 8C.8), the energy is

(8C.16) $E_{one\ state} = \frac{\hbar^2 k^2}{2m}$

223

Then the energy of the shell is,

$$(8C.17) \ dE_{shell} = (N_{electron\ states})(E_{one\ state})$$
$$= (\frac{V}{\pi^2}k^2dk)(\frac{\hbar^2k^2}{2m}) = \frac{\hbar^2V}{2m\pi^2}k^4dk$$

The total energy is just the integration over the Fermi radius,

$$(8C.18) \ E_{total} = \frac{\hbar^2V}{2m\pi^2}\int_0^{k_F}k^4dk$$
$$= \frac{\hbar^2V}{2m\pi^2}\frac{k_F^5}{5} = \frac{\hbar^2V}{10m\pi^2}(3\frac{Nq}{V}\pi^2)^{5/3}$$

Where we used equation 8C.12 in the last step. Expressing the total energy in terms of the volume,

$$(8C.19) \ E_{total} = \frac{\hbar^2(3\pi^2Nq)^{5/3}}{10m\pi^2}V^{-2/3}$$

We see that this formula is analogous the internal thermal energy. We take the derivartive with respect to the volume:

$$(8C.20) \ dE_{total} = -\frac{2}{3}\frac{\hbar^2(3\pi^2Nq)^{\frac{5}{3}}}{10m\pi^2}V^{-\frac{5}{3}}dV$$
$$= -\frac{2}{3}E_{total}\frac{dV}{V}$$

We can link this to equation 4B.2, rewritten below,

(8C.21) dW = −PdV

Equating equations 8C.20 and 8C.21, we get

(8C.22) $P = \dfrac{2}{3}\dfrac{E_{total}}{V}$

This pressure is positive, and acting outward. Its nature is not from electric repulsion between the electrons, and neither from thermal factors but mainly due to quantum considerations that electrons obey the Pauli Exclusion Principle, which is itself a result of the anti-symmetric requirement of the wave function (sec. 7G).

8D. General Case of a Particle Occupying n Quantum States with d_n Degeneracies

Degeneracy simply means that two or more states give the same energy upon a measurement. Consider three identical particles in the 1-D infinite well (section 8B). The energy is (equation 8b.7) reproduced below for those three particles,

(8D.1) $E_{n_A n_B n_C} = \dfrac{\hbar^2 \pi^2}{2ma^2} (n_A^2 + n_B^2 + n_C^2)$

For purposes of illustration, suppose the energy is,

(8D.2) $E_{n_A n_B n_C} = \dfrac{\hbar^2 \pi^2}{2ma^2} \, 363,$

Where we set the sum $n_A^2 + n_B^2 + n_C^2 = 363$

The number 363 is purposely chosen as it contains several combinations of 3 factors whose sum of the squares gives that number. We will label these energy quantum states as (n_A, n_B, n_C). Let's examine each of these combinations,

(i) $363 = 11^2 + 11^2 + 11^2$. The configuration is then labelled as:

$(11,11,11) \rightarrow$ this means all three particles are in the same quantum state (ψ_{11}) and $N_{11} = 3$.

(ii) $363 = 1^2 + 1^2 + 19^2$. In this configuration, there are three possible combinations:

$(1,1,19) \rightarrow$ The 1st and 2nd particles are in state 1, the 3rd in state 19.

$(1,19,1) \rightarrow$ The 1st and 3rd particles are in state 1, the 2nd is in state 19.

$(19,1,1) \rightarrow$ The 1st is in state 19, the 2nd and 3rd are in state 1.

(iii) $363 = 5^2 + 13^2 + 13^2$. Again in this configuration, there are three possible combinations:

$(5,13,13) \rightarrow$ The 1st particle is in state 5, the 2nd and 3rd are in state 13.

$(13,13,5) \rightarrow$ The 1st and 2nd particles are in state 13, the 3rd in state 5.

$(13,5,13) \rightarrow$ The 1st and 3rd particles are in state 13, the 2nd is in state 5.

(iv) $363 = 5^2 + 7^2 + 17^2$. Here in this configuration, there are 6 possible combinations:

$(5,7,17) \rightarrow$ The 1st particle is in state 5, the 2nd in state 7 and the 3rd is in state 17.

$(5,17,5) \rightarrow$ The 1st particle is in state 5, the 2nd particle in state 17, and the 3rd in state 5.

$(7,5,17) \rightarrow$ The 1st particle is in state 7, the 2nd in state 5 and the 3rd is in state 17.

$(7,17,5) \rightarrow$ The 1st particle is in state 7, the 2nd in state 17 and the 3rd is in state 5.

(17,5,7) → The 1st particle is in state 17, the 2nd in state 5 and the 3rd is in state 7.
(17,7,5) → The 1st particle is in state 17, the 2nd in state 7 and the 3rd is in state 5.

Note: there are 4 types of configuration, and 13 (= 1+3+3+6) combinations. The most probable configuration is the last one as it can be achieved in six possible combinations. But the name of the game is to calculate the probabilities.

What is the probability P_n of a particle being in state E_n?

Let's do a couple of examples:

(a) What is the probability of a particle being in the quantum energy state E_1? That state occurs in configuration (ii). There are 3 combinations out of 13 (3/13). In that configuration, the probability of getting E_1 is 2/3. Therefore, $P_1 = \left(\frac{3}{13}\right) \times \left(\frac{2}{3}\right) = \frac{2}{13}$

(b) What about E_5? That state occurs in configurations (iii) and (iv). Let's tackle one configuration at a times.

In configuration (iii), we have 3 combinations out of 13 (3/13). But this time, getting E_5 is 1/3 → $\left(\frac{3}{13}\right) \times \left(\frac{1}{3}\right) = \frac{1}{13}$.

In configuration (iv), we have 6 combinations out of 13 (6/13). Getting E_5 is 1/3 → $\left(\frac{6}{13}\right) \times \left(\frac{1}{3}\right) = \frac{2}{13}$.
Altogether we get $P_5 = \frac{1}{13} + \frac{2}{13} = \frac{3}{13}$
If we carry on in this manner, we get $P_7 = \frac{2}{13}$, $P_{11} = \frac{1}{13}$,

$$P_{13} = \frac{2}{13}, P_{17} = \frac{2}{13}, P_{19} = \frac{1}{13}.$$

Before we proceed to the general case, some comments:

(1) Note that the sum of all the probabilities is one, that is,

$$P_1 + P_5 + P_7 + P_{11} + P_{13} + P_{17} + P_{19}$$
$$= \frac{2}{13} + \frac{3}{13} + \frac{2}{13} + \frac{1}{13} + \frac{2}{13} + \frac{2}{13} + \frac{1}{13} = 1$$

(2) For Fermions, the anti-symmetry of the wave function (sec. 7G) tells us that no two particles can occupy the same states. In configuration (i), all three particles are in the same state, so that is not possible. Similarly with two particles in one state, both configurations (ii) and (iii) are not possible. Only configuration (iv) in which all three particles occupy different states is possible for fermions. In each combination, $P_5 = \frac{1}{3}$; $P_7 = \frac{1}{3}$; $P_{17} = \frac{1}{3}$. Again, the sum of all the probabilities is one.

(3) For bosons, we must look at the symmetric property of the wave function (sec. 7G). In configuration (i) we have only one state $\rightarrow \psi_{11}\psi_{11}\psi_{11}$, which obviously is symmetric under exchange of the particles. So in effect we have one state. For configuration (ii), we have,

$$\psi_1\psi_1\psi_{19} \sim \psi_1\psi_{19}\psi_1 \sim \psi_{19}\psi_1\psi_1$$

Again we have only one state. Ditto for configuration (iii). For configuration (iv), we also have only one state:

$$\psi_5\psi_7\psi_{17} \sim \psi_5\psi_{17}\psi_7 \sim \psi_7\psi_5\psi_{17}$$
$$\sim \psi_7\psi_{17}\psi_5 \sim \psi_{17}\psi_5\psi_7 \sim \psi_{17}\psi_7\psi_5$$

In all we have four states. We can carry a similar calculation. In example (a) $P_1 = \left(\frac{1}{4}\right) \times \left(\frac{2}{3}\right) = \frac{1}{6}$. In example (b) $P_5 = \left(\frac{1}{4}\right) \times \left(\frac{1}{3}\right) + \left(\frac{1}{4}\right) \times \left(\frac{1}{3}\right) = \frac{1}{6}$... and so on. No doubt the total probability is one.

And now we move on to the general case: suppose we put N particles in configurations N_1, N_2, N_3..., that is, we have N_1 particles in state E_1, N_2 particles in state E_2, N_3 particles in state E_3... and so on. Se want to know: in how many ways this can be achieved. The answer $Q(N_1, N_2, N_3...)$ will depend on whether we have distinguishable particles, or identical particles like fermions or bosons. We will do all three cases separately.

First Case: Distinguishable Particles

It's easier to think of the energy states as boxes:
$E_1 \rightarrow$ box 1, $E_2 \rightarrow$ box 2, $E_3 \rightarrow$ box 3 ...

From the N available candidates, how many ways can we put N_1 into box 1?

N chooses N_1 is the binomial coefficient:

(8D.3) $\quad \begin{pmatrix} N \\ N_1 \end{pmatrix} = \dfrac{N!}{N_1!(N-N_1)!}$ (Appendix C, equation CF.1)

Quick proof:
(i) There are N ways to pick the first particle, N −1 for the second and so on,
$$N(N-1)(N-2) ... (N-N_1+1) = \frac{N!}{(N-N_1)!}$$

229

(ii) Now this counts the different ways of picking $N_1!$ permutations of the N_1 particles. We don't care if we pick particle number 25 on the first draw or on the 22nd draw. Hence we divide by $N_1!$

QED

Next we must calculate the different ways the N_1 particles can be arranged within box 1. Say each particle has d_1 distinct choices. Therefore we get $(d_1)^{N_1}$ possibilities in all.

The total number of ways to put N_1 particles, each having d_1 distinct choices in box 1 is,

$$\frac{N!}{N_1!(N-N_1)!}(d_1)^{N_1}$$

Similarly for box 2,

$$\frac{(N-N_1)!}{N_2!(N-N_1-N_2)!}(d_2)^{N_2}$$

It follows that,

(8D.4) $Q(N_1, N_2, N_3 \ldots)$

$$= \frac{N!}{N_1!(N-N_1)!}(d_1)^{N_1}\frac{(N-N_1)!}{N_2!(N-N_1-N_2)!}(d_2)^{N_2}\frac{(N-N_1-N_2)!}{N_3!(N-N_1-N_2-N_3)!}(d_3)^{N_3}\ldots$$

$$= N!\frac{(d_1)^{N_1}(d_2)^{N_2}(d_3)^{N_3}\ldots}{N_1!N_2!N_3!\ldots}$$

$$= N!\prod_{n=1}^{\infty}\frac{(d_n)^{N_n}}{N_n!}$$

Where $\prod_{n=1}^{N}f_n = f_1 \cdot f_2 \cdot f_3 \cdot \ldots f_N$

Second Case: Identical Fermions

Recall in the above example, the fermions could only be placed in configuration (iv) as only one particle can occupy any given state. So the question is reduced to: in the nth box, d_n chooses N_n. From equation 8D.3, we get

$$\binom{d_n}{N_n} = \frac{d_n!}{N_n!\,(d_n - N_n)!}$$

Since there is only one box, equation 8D.4 becomes,

(8D.5) $Q(N_1, N_2, N_3 \ldots) = \prod_{n=1}^{\infty} \frac{d_n!}{N_n!(d_n - N_n)!}$

Third Case: Identical Bosons

Again recall the above example, for the bosons, in each configuration there was only one state, since there are no restrictions on the number of particle that can share that given state. So consider the specific example with $d_n = 5$ and $N_n = 7$. Let dots represent particles and crosses represent partitions:

●● x ● x ●● x ● x ●

This would be the case in which we have two particles in the first state, one in the second, two in the third, one in the fourth and one in the fifth. Note there are N_n dots and $d_n - 1$ crosses. There are $N_n + d_n - 1$ different ways of arranging them in the nth box:

$$\binom{N_n + d_n - 1}{N_n} = \frac{(N_n + d_n - 1)!}{N_n!\,(d_n - 1)!}$$

With only one box this gives,

(8D.6) $\quad Q(N_1, N_2, N_3 \ldots) = \prod_{n=1}^{\infty} \frac{(N_n + d_n - 1)!}{N_n!\,(d_n - 1)!}$

8E. Quantum Mechanical Statistics Distributions

In this part we will derive the Maxwell-Boltzmann distribution for distinguishable particles; the Fermi-Dirac distribution for identical fermions; and the Bose-Einstein distribution for identical bosons.

Given a total energy state E and a given total particle number N, we want to know what is the most probable configuration $(N_1, N_2, N_3 \ldots)$. In other words, we are looking for the particular configuration in which $Q(N_1, N_2, N_3 \ldots)$ is a maximum subject to the constraints,

(8E.1) $\quad \sum_{n=1}^{\infty} N_n = N \quad \rightarrow \quad N - \sum_{n=1}^{\infty} N_n = 0$

And

(8E.2) $\quad \sum_{n=1}^{\infty} N_n E_n = E \quad \rightarrow \quad E - \sum_{n=1}^{\infty} N_n E_n = 0$

To find the maximum of a function we use the Lagrange multiplier (see appendix F). In this case a function $F(x_1, x_2, x_3 \ldots)$ is subject to constraints $f_1(x_1, x_2, x_3 \ldots) = 0$ and $f_2(x_1, x_2, x_3 \ldots) = 0$ and so on. We set

(8E.3) $\quad G(x_1, x_2, x_3 \ldots \lambda_1, \lambda_2 \ldots) \equiv F + \lambda_1 f_1 + \lambda_2 f_2 + \cdots$

With all of its derivatives equal to zero,

(8E.4) $\dfrac{\partial G}{\partial x_n} = 0$; $\dfrac{\partial F}{\partial \lambda_n} = 0$ $n = 1,2,3 \dots$

Now $Q(N_1, N_2, N_3 \dots)$ is a product so it's best to work with the logarithm of Q, $\{F \equiv \ln Q(N_1, N_2, N_3 \dots)\}$, which turns its products into sums. Equation 8E.3 becomes,

(8E.6) $G \equiv \ln(Q) + \alpha(N - \sum_{n=1}^{\infty} N_n)$
$$+ \beta(E - \sum_{n=1}^{\infty} N_n E_n)$$

Where the Lagrange multipliers are: $\alpha = \lambda_1$ and $\beta = \lambda_2$.

Now we will consider all three cases.

First Case: Distinguishable Particles

In this case, Q takes the form of equation 8D.4, reproduced below:

(8E.7) $Q(N_1, N_2, N_3 \dots) = N! \prod_{n=1}^{\infty} \dfrac{(d_n)^{N_n}}{N_n!}$

Substitute that into equation 8E.6 we get,

(8E.8) $G \equiv \ln(N!) + \sum_{n=1}^{\infty} \{N_n \ln(d_n) - \ln(N_n!)\}$
$$+ \alpha(N - \sum_{n=1}^{\infty} N_n) + \beta(E - \sum_{n=1}^{\infty} N_n E_n)$$

Since N_n is a very large number, in order to deal with the logarithm of a factorial, which is in the sum, we can use Stirling's approximation formulation (see appendix G).

(8E.9) $\ln(N_n!) \approx N_n \ln(N_n) - N_n$

(8E.10) An important result is,

$$\frac{\partial \ln(N_n!)}{\partial N_n} = \ln(N_n) + N_n \left(\frac{1}{N_n}\right) - 1 = \ln(N_n)$$

Now we take the derivative of equation 8E.8 with respect to N_n,

(8E.11) $\dfrac{\partial G}{\partial N_n} = 0$

$$= \ln d_n - \ln(N_n) - \alpha - \beta E_n$$

Rearranging,

$$\rightarrow \ln(N_n) = \ln d_n - (\alpha + \beta E_n)$$

Or,

(8E.12) $\quad N_n = d_n e^{-(\alpha + \beta E_n)}$

<u>Second Case</u>: Identical Fermions

For fermions, we use equation 8D.5, reproduced below,

(8E.13) $\quad Q(N_1, N_2, N_3 \dots) = \prod_{n=1}^{\infty} \dfrac{d_n!}{N_n!(d_n - N_n)!}$

Substitute that into equation 8E.6 we get,

(8E.14)
$$G = \sum_{n=1}^{\infty} \{\ln(d_n!) - (\ln(N_n!) + \ln((d_n - N_n)!)\} \\ + \alpha(N - \sum_{n=1}^{\infty} N_n) + \beta(E - \sum_{n=1}^{\infty} N_n E_n)$$

Since $d_n \gg N_n \rightarrow d_n - N_n$ is also very large, so we can apply Stirling's approximation formula to both brackets in the sum. Again taking the derivative of equation 8E.14 with respect to N_n, using equation 8E.10,

(8E.15) $\dfrac{\partial G}{\partial N_n} = 0$

$$= -\ln(N_n) + \ln(d_n - N_n) - \alpha - \beta E_n$$

Rearranging,

$$\rightarrow \ln(N_n) = \ln(d_n - N_n) - \alpha - \beta E_n$$

Or,

$$\rightarrow N_n = (d_n - N_n)e^{-(\alpha + \beta E_n)}$$

$$\rightarrow N_n e^{(\alpha + \beta E_n)} = (d_n - N_n)$$

$$\rightarrow N_n \left(e^{(\alpha + \beta E_n)} + 1\right) = d_n$$

(8E.16) $N_n = \dfrac{d_n}{e^{(\alpha + \beta E_n)} + 1}$

Third Case: Identical Bosons

In this case we use equation 8D.6, reproduced below,

(8E.17) $Q(N_1, N_2, N_3 \dots) = \prod_{n=1}^{\infty} \dfrac{(N_n + d_n - 1)!}{N_n!(d_n - 1)!}$

We repeat our steps: substitute that into 8E.6, take the derivative with respect to N_n, using equation 8E.10,

(8E.18) $\dfrac{\partial G}{\partial N_n} = 0$

$$= \ln(N_n + d_n - 1) - \ln(N_n) - \alpha - \beta E_n$$

Rearranging,

$\rightarrow \ln(N_n) = \ln(N_n + d_n - 1) - \alpha - \beta E_n$

Or,

$\rightarrow N_n = (N_n + d_n - 1)e^{-(\alpha + \beta E_n)}$

$\rightarrow N_n e^{(\alpha + \beta E_n)} = (N_n + d_n - 1)$

$\rightarrow N_n \left(e^{(\alpha + \beta E_n)} - 1\right) = d_n - 1$

(8E.19) $N_n = \dfrac{d_n}{\left(e^{(\alpha + \beta E_n)} - 1\right)}$

Where we dropped the number 1 in the numerator to be consistent with fermions.

Our next and final step is to determine the physical meaning of the Lagrange multipliers α and β. To do that we will use the results in section 8C for a particle in a solid box. In particular equation 8C.8 for the allowed quantized energy, reproduced below,

(8E.20) $E_k = \dfrac{\hbar^2 k^2}{2m}$

Where k is the magnitude of the wave vector (equation 8C.9) $k = (k_x, k_y, k_z) = (\dfrac{n\pi}{l_x}, \dfrac{n\pi}{l_y}, \dfrac{n\pi}{l_z})$.

Also, our "box" now becomes the octant of the spherical shell (Fig. 8C.2). The "degeneracy", that is, the number of states in that box is (equation 8C.15), where we now consider the 2 electrons to be in different states,

(8E.21) $N_{electron\ states} \equiv d_k = \dfrac{V_{shell}}{V_{quantum\ state}}$

$$= \dfrac{\left(\frac{1}{2}\pi k^2 dk\right)}{\frac{\pi^3}{V}}$$

$$= \dfrac{V}{2\pi^2} k^2 dk$$

Substitute that for the first constraint in the case of distinguishable particles,

(8E.22) $N = \sum_{n=1}^{\infty} N_n$ (8E.1)

$$= \sum_{n=1}^{\infty} d_n e^{-(\alpha + \beta E_n)} \text{ (8E.12)}$$

$$= \dfrac{V}{2\pi^2} e^{-\alpha} \int_0^{\infty} e^{-\beta E_k} k^2 dk \text{ (8E.21)}$$

$$= \dfrac{V}{2\pi^2} e^{-\alpha} \int_0^{\infty} e^{-\beta \frac{\hbar^2 k^2}{2m}} k^2 dk \text{ (8E.20)}$$

To evaluate the integral we use from appendix I, equation IE.2,

(8E.23) $\int_0^{+\infty} x^2 e^{-ax^2} dx = \dfrac{1}{2} \int_{-\infty}^{+\infty} x^2 e^{-ax^2} dx$

$$= \dfrac{1}{4}\sqrt{\dfrac{\pi}{a^3}}$$

Set $a = \beta \dfrac{\hbar^2}{2m}$

(8E.24) $N = \dfrac{1}{4}\dfrac{V}{2\pi^2} e^{-\alpha} \sqrt{\dfrac{\pi}{\left(\beta\frac{\hbar^2}{2m}\right)^3}}$

$$= V e^{-\alpha}\dfrac{\pi^{\frac{1}{2}} 2^{\frac{3}{2}}}{\pi^2\, 2^3}\left(\dfrac{m}{\beta\hbar^2}\right)^{\frac{3}{2}}$$

$$= V e^{-\alpha}\left(\dfrac{m}{2\pi\beta\hbar^2}\right)^{3/2}$$

In terms of α, we get

(8E.25) $\quad e^{-\alpha} = \frac{N}{V}\left(\frac{2\pi\beta\hbar^2}{m}\right)^{3/2}$

From the second constraint we have

(8E.26) $\quad E = \sum_{n=1}^{\infty} N_n E_n \quad$ (8E.2)

$\qquad = \sum_{n=1}^{\infty} d_n e^{-(\alpha+\beta E_n)} E_n$ (8E.12)

$\qquad = \frac{V}{2\pi^2} e^{-\alpha} \int_0^{\infty} e^{-\beta E_k} k^2 E_k dk$ (8E.21)

$\qquad = \frac{V}{2\pi^2} e^{-\alpha} \frac{\hbar^2}{2m} \int_0^{\infty} e^{-\beta\frac{\hbar^2 k^2}{2m}} k^4 dk$ (8E.20)

From appendix I, equation IE.3,

(8E.27) $\quad \int_0^{+\infty} x^4 e^{-ax^2} dx = \frac{1}{2}\int_{-\infty}^{+\infty} x^4 e^{-ax^2} dx$

$\qquad\qquad\qquad\qquad = \frac{3}{8}\pi^{\frac{1}{2}} a^{-\frac{5}{2}}$

Again set $a = \beta\frac{\hbar^2}{2m}$

(8E.28) $\quad E = \frac{V}{2\pi^2} e^{-\alpha} \frac{\hbar^2}{2m} \frac{3}{8} \pi^{\frac{1}{2}}\left(\beta\frac{\hbar^2}{2m}\right)^{-\frac{5}{2}}$

$\qquad = \frac{V}{2\pi^2}\frac{N}{V}\left(\frac{2\pi\beta\hbar^2}{m}\right)^{3/2} \frac{\hbar^2}{2m} \frac{3}{8} \pi^{\frac{1}{2}}\left(\beta\frac{\hbar^2}{2m}\right)^{-\frac{5}{2}}$ (8E.25)

$\qquad = 3N \frac{2^{3/2}\cdot 2^{5/2}}{2\cdot 2\cdot 2^3} \frac{\pi^{3/2}\cdot\pi^{1/2}}{\pi^2} \frac{m^{5/2}}{m^{3/2}\cdot m} \frac{\hbar^3\cdot\hbar^2}{\hbar^5} \frac{\beta^{3/2}}{\beta^{5/2}}$

$\qquad = \frac{3N}{2\beta}$

Comparing this to the classical average kinetic energy of an atom at temperature T (equation 4E.13).

(8E.29) $\quad \frac{E}{N} = \frac{3}{2} k_B T$ where k_B is Boltzmann's constant

We identify,

(8E.30) $\beta = \dfrac{1}{k_B T}$

For the Lagrange multiplier α, we define it in terms of the chemical potential as,

(8E.31) $\mu (T) = - \alpha k_B T$

We define the energy ε as the energy in a particular state, and the number of particles with that energy (we divide by the degeneracy)

(8E.32) $n(\epsilon) = \dfrac{N_n}{d_n}$

For the three cases at hand:

(8E.33a) For distinguishable particles (equation 8E.12):

$$n(\epsilon) = e^{-(\epsilon - \mu)/k_B T}$$

Called the Maxwell- Boltzmann distribution.

(8E.33b) For identical fermions (equation 8E.16):

$$n(\epsilon) = \dfrac{1}{e^{(\epsilon - \mu)/k_B T} + 1}$$

Called the Fermi-Dirac distribution.

(8E.33c) For identical bosons (equation 8E.17):

$$n(\epsilon) = \dfrac{1}{e^{(\epsilon - \mu)/k_B T} - 1}$$

Called the Bose-Einstein distribution.

Note: In the particular case of the Fermi-Dirac distribution (8E.33b), as $T \to 0$, the absolute zero temperature, we have

(8E.34a)
$$e^{(\epsilon - \mu)/k_B T} \to 0, \text{ if } \epsilon < \mu(0)$$
$$e^{(\epsilon - \mu)/k_B T} \to \infty, \text{ if } \epsilon > \mu(0)$$

(8E.34b)
$$n(\epsilon) \to 1, \text{ if } \epsilon < \mu(0)$$
$$n(\epsilon) \to 0, \text{ if } \epsilon > \mu(0)$$

This is again the Pauli Exclusion Principle at work. It tells us that at absolute temperature for non-interacting fermions, all the states are occupied and none are occupied above that energy. This is the Fermi energy ($\mu(0) \equiv E_F$).

8F. Blackbody Radiation

To derive Planck's law of blackbody radiation, we need to recall certain assumptions which were the basis that led Planck to realize that the energy for a blackbody radiating is quantized.

(i) The energy of a single photon is related to its frequency, $E = \hbar\omega$.

(ii) The wave number is related to the frequency $k = \omega/c$.

(iii) The relativistic energy of a photon is given by E= pc (equation 5G.27 with m=0), where p is the momentum, and c is the speed of light. So, E $= \hbar\omega = \hbar k c \rightarrow p = \hbar k$

(v) The number of photons is not conserved. In equation 8E.31, $\alpha \rightarrow 0$. So equation 8E.19 now reads as,

(8F.1) $N_\omega = \dfrac{d_k}{(e^{\hbar\omega/k_B T}-1)}$

For free electrons in a box of volume V, d_k is given by equation 8C.15, reproduced below,

(8F.2) $d_k = \dfrac{V}{\pi^2} k^2 dk = \dfrac{V}{\pi^2} \dfrac{\omega^2}{c^2} d\left(\dfrac{\omega}{c}\right) = \dfrac{V}{\pi^2} \dfrac{\omega^2}{c^3} d\omega$

In the blackbody radiation, one looks for the energy density, which in the frequency range $d\omega$, is taken to be, $\rho(\omega)\, d\omega$. Now, the number of particles per unit volume times the energy of a single photon is,

(8F.3) $\dfrac{N_\omega}{V}\, \hbar\omega = \dfrac{d_k}{(e^{\hbar\omega/k_B T}-1)}\, \dfrac{\hbar\omega}{V}$ (8F.1)

$= \dfrac{1}{(e^{\hbar\omega/k_B T}-1)}\, \dfrac{V}{\pi^2} \dfrac{\omega^2}{c^3}\, d\omega\, \dfrac{\hbar\omega}{V}$ (8F.2)

$= \dfrac{\hbar\omega^3}{\pi^2 c^3 (e^{\hbar\omega/k_B T}-1)}\, d\omega$

(8F.4) Therefore we identify,

$$\rho(\omega) \equiv \dfrac{\hbar\omega^3}{\pi^2 c^3 (e^{\hbar\omega/k_B T}-1)}$$

This is the famous Planck's equation for a blackbody spectrum. Historically, it was believed that the intensity emitted by a blackbody was inversely proportional to the wavelength (dotted line in Fig. 8F.1).

The other lines are for emissions at different temperatures, what was actually observed and corrected by Planck's equation. At smaller wavelength, the known theory at that time predicted that the intensity would be infinite – why it was called the ultraviolet catastrophe. Planck, followed by Einstein, resolved this puzzle with the assumptions listed at the beginning of this section.

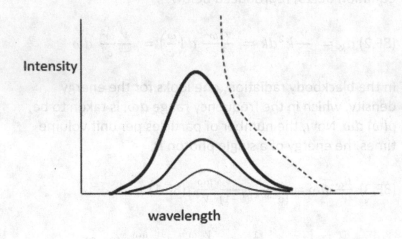

Fig. 8F.1

Chapter 9

Quantum Tunneling

While in chapter 8 we were mainly looking at particles trapped in a potential well, in this chapter we will examine particles trying to overcome a potential barrier.

9A. The Classical Case

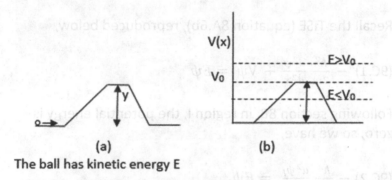

(a)   (b)

The ball has kinetic energy E

Fig. 9A.1

In the classical case, a ball with kinetic energy (Fig. 9A.1a) will overcome a potential energy barrier if its initial energy is greater than the potential energy of the barrier (Fig. 9A.1b). If not, it will bounce back and there is no penetration into the barrier.

9B. The Quantum Case

If the quantum particle has energy E less than the potential barrier, there is a non-zero probability of finding the particle in the classically forbidden region (Ψ_{III} in Fig. 9A.2a).

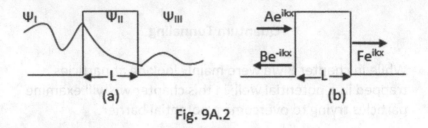

(a)   (b)

Fig. 9A.2

9C. Transmission Coefficient

Recall the TISE (equation 8A.6b), reproduced below,

(9C.1) $-\frac{\hbar^2}{2m}\frac{d^2\psi}{d^2x} + V\psi = E\psi$

Following section 8B, in region I, the potential energy is zero, so we have,

(9C.2) $-\frac{\hbar^2}{2m}\frac{d^2\psi_I}{d^2x} = E\psi_I$

$\rightarrow \frac{d^2\psi_I}{d^2x} = -k_1{}^2\psi_I$, where $k_1 = \frac{\sqrt{2mE}}{\hbar}$

A solution is

(9C.3) $\psi_I = Ae^{i\,k_1 x} + Be^{-ik_1 x}$

Where $Ae^{i\,k_1 x}$ is the incident wave, and $Be^{-ik_1 x}$ is the reflected wave (fig. 9A.2b).

In region II, we have to deal with the potential barrier, so our TISE is,

(9C.4)) $-\frac{\hbar^2}{2m}\frac{d^2\psi_{II}}{d^2x} = (E-V)\psi_{II}$

$\rightarrow \frac{d^2\psi_{II}}{d^2x} = -k_2{}^2\psi_{II}$,where $k_2 = \frac{\sqrt{2m(V-E)}}{\hbar}$

A solution is,

(9C.5) $\psi_{II} = Ce^{k_2x} + De^{-k_2x}$

Also in region III, the potential V= 0, so we have again a similar solution to 9C.2, with different constants,

(9C.6) $\psi_{III} = Fe^{ik_1x} + Ge^{-ik_1x}$

So we have 6 constants to solve for.

Our initial wave was on the left moving to the right: it can penetrate or reflect. But in region III, there is only a transmitted wave moving to the right: so $G = 0$.

To solve for the remaining 5 constants, we apply boundary conditions, that is, the wave function must be continuous, and its derivatives must also be continuous.

On the boundary between regions I and II, with x = 0, we have,

(9C.7) $\psi_I(0) = \psi_{II}(0)$

$\rightarrow A + B = C + D$

Similarly for its derivatives:

(9C.8) $\dfrac{\partial \psi_I(0)}{\partial x} = \dfrac{\partial \psi_{II}(0)}{\partial x}$

$\rightarrow Aik_1 - Bik_1 = Ck_2 - Dk_2$

We use the same reasoning at x = L, between regions II and III, that is,

(9C.9) $\psi_{II}(L) = \psi_{III}(L)$

$\rightarrow Ce^{k_2 L} + De^{-k_2 L} = Fe^{ik_1 L}$

For its derivatives,

(9C.10) $\dfrac{\partial \psi_{II}(L)}{\partial x} = \dfrac{\partial \psi_{III}(L)}{\partial x}$

$\rightarrow Ck_2 e^{k_2 L} - Dk_2 e^{-k_2 L} = Fik_1 e^{ik_1 L}$

The next step is to combine equations 9C.7 and 9C.8. First we multiply equation 9C.7 by ik_1. And then we add equation 9C.8. We get,

(9C.11) $Aik_1 + Bik_1 = Cik_1 + Dik_1$
$\qquad\quad Aik_1 - Bik_1 = Ck_2 - Dk_2$

$\rightarrow \qquad 2Aik_1 = C(ik_1 + k_2) + D(ik_1 - k_2)$

Divide by $2ik_1$.

(9C.12) $A = \dfrac{C(ik_1 + k_2) + D(ik_1 - k_2)}{2ik_1}$

We also combine equations 9C.9 and 9C.10. First we multiply equation 9C.9 by k_2. And then we add equation 9C.10. We get,

(9C.13) $Ck_2e^{k_2L} + Dk_2e^{-k_2L} = Fk_2e^{ik_1L}$

$\qquad\quad Ck_2e^{k_2L} - Dk_2e^{-k_2L} = Fik_1e^{ik_1L}$

$\rightarrow \qquad 2Ck_2e^{k_2L} = F(ik_1 + k_2)e^{ik_1L}$

Divide both sides by $2k_2e^{k_2L}$.

(9C.14) $C = \dfrac{F(ik_1+k_2)e^{ik_1L}}{2k_2e^{k_2L}}$

To get D we repeat 9C.13, except we subtract, that is,

(9C.15) $Ck_2e^{k_2L} + Dk_2e^{-k_2L} = Fk_2e^{ik_1L}$

$\qquad\quad -Ck_2e^{k_2L} + Dk_2e^{-k_2L} = -Fik_1e^{ik_1L}$

And solve for D:

(9C.16) $D = \dfrac{F(k_2-ik_1)e^{ik_1L}}{2k_2e^{-k_2L}}$

Now that we have expressions for constants C and D, we can substitute those into equation 9C.12:

(9.C.17) $A = \dfrac{\left[\dfrac{F(ik_1+k_2)e^{ik_1L}}{2k_2e^{k_2L}}\right](ik_1+k_2) + \left[\dfrac{F(k_2-ik_1)e^{ik_1L}}{2k_2e^{-k_2L}}\right](ik_1-k_2)}{2ik_1}$

$= F\left\{\left[\dfrac{(ik_1+k_2)e^{ik_1L}}{4ik_1k_2e^{k_2L}}\right](ik_1+k_2) + \left[\dfrac{(k_2-ik_1)e^{ik_1L}}{4ik_1k_2e^{-k_2L}}\right](ik_1-k_2)\right\}$

Put the two terms under a common denominator:

(9C.18)
$$\frac{A}{F} = \frac{(ik_1+k_2)e^{ik_1L}(ik_1+k_2)e^{-k_2L}+(k_2-ik_1)e^{ik_1L}(ik_1-k_2)e^{k_2L}}{4ik_1k_2}$$

Factor out e^{ik_1L}.

(9C.19)

$$\frac{A}{F} = \frac{e^{ik_1L}}{4ik_1k_2} [(ik_1+k_2)(ik_1+k_2)e^{-k_2L}$$
$$+ (k_2-ik_1)(ik_1-k_2)e^{k_2L}]$$

We can rearrange the terms in the square bracket:

(9C.20a) $(ik_1+k_2)(ik_1+k_2) = \left(k_2{}^2 - k_1{}^2\right) + 2ik_1k_2$
(9C.20b) $(k_2-ik_1)(ik_1-k_2) = -\left(k_2{}^2 - k_1{}^2\right) + 2ik_1k_2$

This gives

(9C.21) $\dfrac{A}{F} = \dfrac{e^{ik_1L}}{4ik_1k_2}[\left(k_2{}^2 - k_1{}^2\right)e^{-k_2L} + 2ik_1k_2e^{-k_2L}$
$$-\left(k_2{}^2 - k_1{}^2\right)e^{k_2L} + 2ik_1k_2e^{k_2L}]$$

$$= \frac{e^{ik_1L}}{4ik_1k_2}[\left(k_2{}^2 - k_1{}^2\right)(e^{-k_2L} - e^{k_2L})$$
$$+2ik_1k_2(e^{-k_2L} + e^{k_2L})]$$

From Appendix C, equations CE.9 and CE.10,

(9C.22a) $2\sinh k_2L = (e^{-k_2L} - e^{k_2L})$ (CE.9−CE.10)

(9C.22b) $2\cosh k_2L = (e^{-k_2L} + e^{k_2L})$ (CE.9+CE.10)

Substitute, we obtain for the ratio A/F:

(9C.22c)
$$\frac{A}{F} = \frac{e^{ik_1L}}{4ik_1k_2}[2(k_2^2 - k_1^2)\sinh k_2L + 4ik_1k_2\cosh k_2L]$$

What we are interested is the transmission coefficient (Fig. 9A.2b):

$$(9.C23) \quad T = \frac{|\Psi_{transmitted}|^2}{|\Psi_{incident}|^2} = \frac{|F|^2}{|A|^2} = \frac{FF^*}{AA^*}$$

So we need to take the complex conjugate of equation · 9C.22c. We get,

(9C.24)
$$\frac{A^*}{F^*} = -\frac{e^{-ik_1L}}{4ik_1k_2}[2(k_2^2 - k_1^2)\sinh k_2L$$
$$- 4ik_1k_2\cosh k_2L]$$

Multiplying equation 9C.22 with equation 9C.24.

$$(9C.25) \quad \frac{|A|^2}{|F|^2} = \frac{AA^*}{FF^*} = \frac{1}{16k_1^2k_2^2}[4(k_2^2 - k_1^2)^2\sinh^2(k_2L)$$
$$+16k_1^2k_2^2\cosh^2(k_2L)]$$

From appendix C, equation CE.5,

$$\cosh^2\theta = 1 + \sinh^2\theta$$

$$\rightarrow \frac{|A|^2}{|F|^2} = \frac{AA^*}{FF^*} = \frac{1}{16k_1^2k_2^2}[4(k_2^2 - k_1^2)^2\sinh^2(k_2L)$$
$$+16k_1^2k_2^2\{1 + \sinh^2(k_2L)\}]$$
$$= \frac{1}{4k_1^2k_2^2}[(k_2^4 - 2k_1^2k_2^2 + k_1^4 + 4k_1^2k_2^2)\sinh^2(k_2L)$$
$$+4k_1^2k_2^2]$$

$$= \frac{1}{4k_1{}^2 k_2{}^2} \left[\left(k_2{}^2 + k_1{}^2 \right)^2 \sinh^2(k_2 L) + 4k_1{}^2 k_2{}^2 \right]$$

$$= 1 + \frac{\left(k_2{}^2 + k_1{}^2 \right)^2 \sinh^2(k_2 L)}{4k_1{}^2 k_2{}^2}$$

Now equation 9C.23 is the reciprocal,

(9C.26) $\quad T = \dfrac{|F|^2}{|A|^2} = \dfrac{1}{1 + \dfrac{\left(k_2{}^2 + k_1{}^2 \right)^2 \sinh^2(k_2 L)}{4k_1{}^2 k_2{}^2}}$

While Classical physics predicts that if the energy of the particle is less than the potential barrier (E<V) there is no transmission, we see that the transmission coefficient is greater than zero, a phenomenon only explained by Quantum Mechanics. Quantum tunneling plays an essential role in such cases as nuclear fusion, in diodes, quantum computing and the main sequence in stars.

Chapter 10

Relativistic Quantum Mechanics

10A. The Klein-Gordon Equation

We start with the non-relativistic Hamiltonian for a free particle,

(10A.1) $H = \dfrac{p^2}{2m}$

The above is obtained from equation 2B.2 with V = 0, and
$T = \frac{1}{2} mv^2 = \dfrac{m^2v^2}{2m} = \dfrac{p^2}{2m}$

Recall the Schrödinger equation (7C.9):

(10A.2) $i\hbar \dfrac{\partial}{\partial t} \phi = H\phi$

And equation (7C.7) extended to 3-D,

(10A.3) $\mathbf{p} = -i\hbar\nabla$

Altogether we get the non-relativistic Schrödinger equation,

(10A.4) $-\dfrac{\hbar^2}{2m} \nabla^2 \phi = i\hbar \dfrac{\partial}{\partial t} \phi$

One fault to this equation is that it treats space and time asymmetrical – 1st order in time, 2nd order in space. What Is required is an equation that puts space and time to the same order. At the same time, we want our equation to be compatible with SR. So we consider the Hamiltonian (equation 5G.27),

(10A.5) $H^2 = p^2 c^2 + m^2 c^4$

We can do that by squaring the RHS of equation 10A.2:

(10A.6) $H = i\hbar \frac{\partial}{\partial t} \rightarrow H^2 = (i\hbar \frac{\partial}{\partial t})^2$

Putting this altogether, we substitute equations 10A.3 and 10A.6 into 10A.5 we get,

(10A.7) $(i\hbar \frac{\partial}{\partial t})^2 \phi = ((-i\hbar\nabla)^2 c^2 + m^2 c^4)\phi$

A standard procedure is to set $\hbar = c = 1$. We get,

(10A.8) $\left(\frac{\partial^2}{\partial t^2} - \nabla^2 + m^2 \right) \phi = 0$

We define (appendix E, equation EA.16),

(10A.9) $\partial^2 = \partial_\mu \partial^\mu = \frac{\partial^2}{\partial t^2} - \nabla^2$

The Klein-Gordon equation in its standard form now reads as,

(10A.10a) $(\partial^2 + m^2)\phi = 0$

(10A.10b) Or, $(\Box + m^2)\phi = 0$

Where $\Box \equiv \partial^2 = \partial_\mu \partial^\mu$ is the D'Alembertian operator.

The solutions are plane waves of the type:

(10A.11) $\phi \sim e^{i(k \cdot x - \omega t)}$

Where $E = \omega$, $p = k$ and $\omega^2 = k^2 + m^2$

Comments

(i) The energy solutions, $E = \pm\sqrt{p^2 + m^2}$, contain both positive and negative quantities. For a free particle, the negative solution is unphysical.

(ii) It also contains negative probabilities. To see this, we define the probability current as,

$$J = -i\phi^* \frac{\partial \phi}{\partial x} + i\phi \frac{\partial \phi^*}{\partial x}$$

The probability density in QM is defined as,

$$\rho = |\phi|^2 = \phi^*\phi$$

These two quantities are governed by the conservation equation,

$$\frac{\partial \rho}{\partial t} + \frac{\partial J}{\partial x} = 0$$

Substitute the definition of J, we get,

$$\rightarrow \frac{\partial \rho}{\partial t} = -\frac{\partial}{\partial x}\left(-i\phi^* \frac{\partial \phi}{\partial x} + i\phi \frac{\partial \phi^*}{\partial x}\right)$$

$$= i\left(\frac{\partial \phi^*}{\partial x}\frac{\partial \phi}{\partial x} + \phi^* \frac{\partial^2 \phi}{\partial^2 x} - \frac{\partial \phi}{\partial x}\frac{\partial \phi^*}{\partial x} - \phi \frac{\partial^2 \phi^*}{\partial^2 x}\right)$$

$$= i\phi^* \frac{\partial^2 \phi}{\partial^2 x} - i\phi \frac{\partial^2 \phi^*}{\partial^2 x} \quad (1^{st} \text{ and } 3^{rd} \text{ terms cancel})$$

Using the K-G equation (10A.8) in 1-D, we have

$$\frac{\partial^2 \phi}{\partial^2 x} = \frac{\partial^2 \phi}{\partial^2 t} + m^2 \phi \quad \text{and} \quad \frac{\partial^2 \phi^*}{\partial^2 x} = \frac{\partial^2 \phi^*}{\partial^2 t} + m^2 \phi^*$$

Therefore,

$$\frac{\partial \rho}{\partial t} = i\phi^* \left(\frac{\partial^2 \phi}{\partial^2 t} + m^2 \phi \right) - i\phi \left(\frac{\partial^2 \phi^*}{\partial^2 t} + m^2 \phi^* \right)$$

$$= i\phi^* \frac{\partial^2 \phi}{\partial^2 t} - i\phi \frac{\partial^2 \phi^*}{\partial^2 t}$$

The equation is satisfied if,

$$\rho = i\phi^* \frac{\partial \phi}{\partial t} - i\phi \frac{\partial \phi^*}{\partial t}$$

Recall equation 10A.11,

$$\phi \sim e^{i(\boldsymbol{p} \cdot \boldsymbol{x} - Et)}$$

Substituting we get,

$$\rho \sim i(-iE) - i(iE) = 2E$$

But $E = \pm\sqrt{p^2 + m^2}$. For negative energies, the probability density ρ would also be negative. In probability theory, a negative probability has no meaning. Feynman resolved this issue by considering anti-particles with positive energies. The meaning of this in an interaction is the followings: the absorption of a particle is equivalent to the emission of an anti-particle.

(iii) In QFT, the function φ is a scalar field, associated with a spin-zero particle. For an electron with spin ½, the K-G equation is not the candidate.

(iv) The Lagrangian that will yield the K-G equation is,

(10A.12) $\mathcal{L} = ½ (\partial_\mu \phi)^2 - ½m^2\phi^2$

Proof: The Euler-Lagrange equations for fields (equation 2C.7) are, reproduced below:

(10A.13) $\partial_\mu(\partial\mathcal{L}/\partial(\partial_\mu\phi)) - \partial\mathcal{L}/\partial\phi = 0$

(i) Do the first term: in equation 10A.12, take the derivative $\partial\mathcal{L}/\partial(\partial_\mu\phi) = \partial_\mu\phi$
Take the second derivative: $\partial_\mu(\partial\mathcal{L}/\partial(\partial_\mu\phi)) = \partial^2\phi$

(ii) Do the second term: $\partial\mathcal{L}/\partial\phi = -m^2\phi$

(iii) Subtract: $\partial^2\phi + m^2\phi = 0$ (equation 10A.13)

QED

A complete general solution to the K-G equation (10A.10) will then be:

(10A.14) $\phi(\mathbf{x}, t) = \int \frac{d^3k}{(2\pi)^{3/2}(2E_p)^{1/2}} (a^\dagger{}_p e^{ipx} + b_p e^{-ipx})$

Where $a^\dagger{}_p$ is a creation operator for a particle, b_p is an annihilator operator for an anti-particle, and E_p is the energy of a free particle used as a normalization factor. More to be said on this in chapter 11.

10B. The Dirac Equation

Dirac sought an equation of the first order. He proposed,

(10B.1) $H\psi = (\boldsymbol{\alpha} \cdot \boldsymbol{p} + \beta m)\psi$

Or $\quad i\hbar\frac{\partial}{\partial t}\psi = (-i\boldsymbol{\alpha} \cdot \boldsymbol{\nabla} + \beta m)\psi$

Where $\boldsymbol{\alpha} = (\alpha_1, \alpha_2, \alpha_3)$, with $\alpha_1, \alpha_2, \alpha_3$ and β as matrices, and ψ is now a column vector. To determine these matrices we rewrite the above equation as,

(10B.2) $\left(i\hbar\frac{\partial}{\partial t} + i\boldsymbol{\alpha} \cdot \boldsymbol{\nabla} - \beta m\right)\psi = 0$

And multiply $i\hbar\frac{\partial}{\partial t} - i\boldsymbol{\alpha} \cdot \boldsymbol{\nabla} + \beta m$ throughout. We get,

$(i\hbar\frac{\partial}{\partial t} - i\boldsymbol{\alpha} \cdot \boldsymbol{\nabla} + \beta m)\left(i\hbar\frac{\partial}{\partial t} + i\boldsymbol{\alpha} \cdot \boldsymbol{\nabla} - \beta m\right)\psi$

$= \left(i\hbar\frac{\partial}{\partial t} - i\sum_i \alpha_i \partial_i + \beta m\right)\left(i\hbar\frac{\partial}{\partial t} + i\sum_j \alpha_j \partial_j - \beta m\right)\psi$

We now set $\hbar = 1$,

(10B.3) $[-\frac{\partial^2}{\partial t^2} + \sum_i \alpha_i^2 \partial_i \partial_i + \sum_{i \neq j}(\alpha_i\alpha_j + \alpha_j\alpha_i)\partial_i \partial_j$
$\qquad + im \sum_i(\alpha_i\beta + \beta\alpha_i)\partial_i - \beta^2 m^2]\psi = 0$

This equation is equal to the K-G equation if and only if,

(10B.4) $\beta^2 = 1$; $\sum_i \alpha_i^2 = 1$;
$\quad \alpha_i\alpha_j + \alpha_j\alpha_i = 0, \text{for i} \neq \text{j} ; \alpha_i\beta + \beta\alpha_i = 0, \text{i} = 1,2,3$

The quantities that satisfy something similar are the Pauli matrices (equation 7F.16), rewritten below,

(10B.5a)

$$\sigma_1 \equiv \begin{pmatrix} 0 & 1 \\ 1 & 0 \end{pmatrix}, \ \sigma_2 \equiv \begin{pmatrix} 0 & -i \\ i & 0 \end{pmatrix}, \sigma_3 \equiv \begin{pmatrix} 1 & 0 \\ 0 & -1 \end{pmatrix}$$

For reasons that will become apparent we add to this,

(10B.5b) $\quad \sigma_0 \equiv \begin{pmatrix} 1 & 0 \\ 0 & 1 \end{pmatrix} \equiv I$

We define the α_i's and β in terms of the Pauli matrices as, (note: there are several representations, the one following is called the chiral representation):

(10B.6) $\alpha^i = \begin{pmatrix} -\sigma^i & 0 \\ 0 & \sigma^i \end{pmatrix}$; $\beta = \begin{pmatrix} 0 & \sigma^0 \\ \sigma^0 & 0 \end{pmatrix}$,

Where $\mathbf{0} = \begin{pmatrix} 0 & 0 \\ 0 & 0 \end{pmatrix}$

Example: $\alpha^1 = \begin{pmatrix} -\sigma^1 & 0 \\ 0 & \sigma^1 \end{pmatrix} = \begin{pmatrix} 0 & -1 & 0 & 0 \\ -1 & 0 & 0 & 0 \\ 0 & 0 & 0 & 1 \\ 0 & 0 & 1 & 0 \end{pmatrix}$

The Dirac matrices are then defined as:

(10B.7) $\gamma^0 = \begin{pmatrix} 0 & \sigma_0 \\ \sigma_0 & 0 \end{pmatrix}$, $\gamma = \begin{pmatrix} 0 & \sigma \\ -\sigma & 0 \end{pmatrix}$

Finally, we define $\sigma^\mu = (I, \sigma)$ and $\bar{\sigma}^\mu = (I, -\sigma)$ to obtain,

(10B.8) $\gamma^\mu = \begin{pmatrix} 0 & \sigma^\mu \\ \bar{\sigma}^\mu & 0 \end{pmatrix}$

Note that the γ's obey an anti-commutation relations:

(10B.9) $\{\gamma^\mu, \gamma^\nu\} = g^{\mu\nu}$

Where the metric tensor is given by

$$g^{\mu\nu} = \begin{pmatrix} 1 & 0 & 0 & 0 \\ 0 & -1 & 0 & 0 \\ 0 & 0 & -1 & 0 \\ 0 & 0 & 0 & -1 \end{pmatrix}$$

This gives the famous Dirac equation.

(10B.10a) $\left(i\gamma^\mu \partial_\mu - m\right)\psi = 0$

This equation is also written with the slash (/) notation, introduced by Feynman,

(10B.10b) $(\not{p} - m)\psi = 0$

Where $\not{p} = i\gamma^\mu \partial_\mu$

Note: since the γ's are 4X4 matrices, for consistency, ψ is a (4X1) column vector.

In "this innocent-looking" equation, there are many outstanding features one can demonstrate. Among them, we will show (a) that the spin of an electron comes out naturally from that equation; (b) that there exists particles (electrons, e⁻) and anti-particles (positrons, e⁺) with the same mass but opposite charges; (c) that leptons (electrons, muons, neutrinos, and so on) can exist in right-handed (R) quantum states and/or left-handed (L) quantum states – this concept goes by the name of chirality; (d) that right-handed particles have a helicity of +1, while for anti-particles, it is the left-handed version that carries a helicity of +1 (we will define helicity later on).

We start by expanding equation 10B.10a,

(10B.11) $(\gamma_0 p^0 - \boldsymbol{\gamma} \cdot \boldsymbol{p} - m)\psi = 0$

Where $p^0 = i\partial_0$ and $\boldsymbol{p} = -i\nabla$

Using equation 10B.7, in which the γ's are expressed as 2X2 matrices, we will also express the four-component wave function into a (2X1) column vector,

(10B.12) $\psi = \begin{pmatrix} \psi_L \\ \psi_R \end{pmatrix}$

From equations 10B.5b, 10B.7 and 10B12, equation 10B.11 becomes,

(10B.13)

$$\left[\begin{pmatrix} 0 & p^0 \\ p^0 & 0 \end{pmatrix} - \begin{pmatrix} 0 & \boldsymbol{\sigma} \cdot \boldsymbol{p} \\ -\boldsymbol{\sigma} \cdot \boldsymbol{p} & 0 \end{pmatrix} - \begin{pmatrix} m & 0 \\ 0 & m \end{pmatrix} \right] \begin{pmatrix} \psi_L \\ \psi_R \end{pmatrix} = 0$$

We get two equations,

(10B.14a) $(p^0 - \boldsymbol{\sigma} \cdot \boldsymbol{p})\psi_R = m\psi_L$
(10B.14b) $(p^0 + \boldsymbol{\sigma} \cdot \boldsymbol{p})\psi_L = m\psi_R$

The interesting part is when we set m = 0, that is, looking at a massless particle, we get

(10B.15a) $(p^0 - \boldsymbol{\sigma} \cdot \boldsymbol{p})\psi_R = 0$
(10B.15b) $(p^0 + \boldsymbol{\sigma} \cdot \boldsymbol{p})\psi_L = 0$

We see that the two-components parts, ψ_R and ψ_L, do not mix. This tells us that there are two kinds of massless particles: right-handed particles described in the lower slots of the wave function, ψ_R; and left-handed particles in the upper slots, ψ_L. We can define a chirality operator,

(10B.16) $\gamma^5 \equiv i\gamma^0\gamma^1\gamma^2\gamma^5$

From equation 10B.7, we can work out this as,

$$i\gamma^0\gamma^1\gamma^2\gamma^5$$
$$= i \begin{pmatrix} 0 & \sigma_0 \\ \sigma_0 & 0 \end{pmatrix} \begin{pmatrix} 0 & \sigma_1 \\ -\sigma_1 & 0 \end{pmatrix} \begin{pmatrix} 0 & \sigma_2 \\ -\sigma_2 & 0 \end{pmatrix} \begin{pmatrix} 0 & \sigma_3 \\ -\sigma_3 & 0 \end{pmatrix}$$

$$= \begin{pmatrix} i\sigma_0\sigma_1\sigma_2\sigma_3 & 0 \\ 0 & -i\sigma_0\sigma_1\sigma_2\sigma_3 \end{pmatrix}$$

Using equations 10B.5a-b, the element

$$i\sigma_0\sigma_1\sigma_2\sigma_3 = i \begin{pmatrix} 1 & 0 \\ 0 & 1 \end{pmatrix} \begin{pmatrix} 0 & 1 \\ 1 & 0 \end{pmatrix} \begin{pmatrix} 0 & -i \\ i & 0 \end{pmatrix} \begin{pmatrix} 1 & 0 \\ 0 & -1 \end{pmatrix}$$
$$= \begin{pmatrix} -1 & 0 \\ 0 & -1 \end{pmatrix} = -\begin{pmatrix} 1 & 0 \\ 0 & 1 \end{pmatrix} = -I$$

Putting these together, we get

(10B.17) $\gamma^5 = \begin{pmatrix} -I & 0 \\ 0 & I \end{pmatrix}$

Applying this operator on the following:

(i) On a left-handed wave function:

$$\gamma^5 \begin{pmatrix} \psi_L \\ 0 \end{pmatrix} = \begin{pmatrix} -I & 0 \\ 0 & I \end{pmatrix} \begin{pmatrix} \psi_L \\ 0 \end{pmatrix} = - \begin{pmatrix} \psi_L \\ 0 \end{pmatrix}$$

This is an eigenvalue equation with eigenvalue -1.

(ii) Similarly, for right-handed wave function:

$$\gamma^5 \begin{pmatrix} 0 \\ \psi_R \end{pmatrix} = \begin{pmatrix} -I & 0 \\ 0 & I \end{pmatrix} \begin{pmatrix} 0 \\ \psi_R \end{pmatrix} = + \begin{pmatrix} 0 \\ \psi_R \end{pmatrix}$$

This is an eigenvalue equation with eigenvalue +1.

So the Dirac equation predicts the existence of left-handed particles with chirality -1, and right-handed particles with chirality +1. Moreover, for massless particles from equation 10A.5, $p^0 \equiv E_p = |p|$. From the two equations 10B.15a-b, we get,

(10B.18a) $\dfrac{\sigma \cdot p}{|p|} \psi_R = \psi_R \quad \rightarrow h\psi_R = \psi_R$

(10B.18b) $\dfrac{\sigma \cdot p}{|p|} \psi_L = -\psi_L \quad \rightarrow h\psi_L = -\psi_L$

Where we define the helicity operator as,

(10B-2.19) $h \equiv \dfrac{\sigma \cdot p}{|p|}$

With eigenvalues of +1 for right-handed particles, and -1 for left-handed particles. This tells us that the helicity and chirality are identical for massless particles. This will not turn to be true in general.

We turn our attention to the negative energy solutions, which we encountered in solving the K-G equations (see equation 10A.10, comment i). Dirac proposed that these states represent anti-particles. So we write $p^0 \rightarrow -|p^0|$ in the two equations 10B.15a-b, we get,

(10B.20a) $(- |p^0| - \boldsymbol{\sigma} \cdot \boldsymbol{p})\psi_R = 0$
(10B.20b) $(- |p^0| + \boldsymbol{\sigma} \cdot \boldsymbol{p})\psi_L = 0$

For anti-particles, this implies,

(10B.21a) $\frac{\boldsymbol{\sigma} \cdot \boldsymbol{p}}{|p|} \psi_R = - \psi_R \quad \rightarrow \quad h\psi_R = -\psi_R$
(10B.21b) $\frac{\boldsymbol{\sigma} \cdot \boldsymbol{p}}{|p|} \psi_L = \psi_L \quad \rightarrow \quad h\psi_L = \psi_L$

Note that this is the other way around when comparing to the particle case. So if a right-handed particle has helicity +1, its right-handed anti-particle will have the -1 helicity. Also, this treatment predicts that particles and anti-particles have the same mass m in the Dirac equation. Furthermore if a system is made up of energy E and charge −q, a system with energy −E and charge +q will also satisfy the Dirac equation.

In summary so far we have a wave function that contains two slots which behave differently from each other. The two components $\psi_R \; and \; \psi_R$ are known as Weyl spinors.

It is customary to describe Fermi particles with the positive frequency, and negative frequency for Fermi anti-particles. That is,

(10B.22a) $u(p)e^{-ip\cdot x} = \begin{pmatrix} u_L(p) \\ u_R(p) \end{pmatrix} e^{-ip\cdot x}$, for particles

(10B.22b) $v(p)e^{ip\cdot x} = \begin{pmatrix} v_L(p) \\ v_R(p) \end{pmatrix} e^{ip\cdot x}$, for anti-particles

The components $u(p)$ and $v(p)$ are momentum space spinors satisfying the momentum Dirac equation (10B.10b):

(10B.23a) $(\not{p} - m)\, u(p) = 0$
(10B.23b) $(-\not{p} - m)\, v(p) = 0$

To find the solutions of these equations, we first consider particles at rest, $p^\mu = (m, 0)$. $\left[p^\mu = (p^0, p^i), and\ p^0 = E = m\ with\ c = 1, and\ p^i = 0\ for\ a\ particle\ at\ rest. \right]$. For the first equation 10B.23a we have,

(10B.24) $(\gamma_0 p^0 - m) \begin{pmatrix} u_L(p^0) \\ u_R(p^0) \end{pmatrix} = 0$

$$\left[\begin{pmatrix} 0 & I \\ I & 0 \end{pmatrix} m - m \begin{pmatrix} I & 0 \\ 0 & I \end{pmatrix} \right] \begin{pmatrix} u_L(p^0) \\ u_R(p^0) \end{pmatrix} = 0$$

$$\begin{pmatrix} -m & m \\ m & -m \end{pmatrix} \begin{pmatrix} u_L(p^0) \\ u_R(p^0) \end{pmatrix} = 0$$

Solutions are of the type,

(10B.25) $u(p^0) \equiv \begin{pmatrix} u_L(p^0) \\ u_R(p^0) \end{pmatrix} = \sqrt{m} \begin{pmatrix} \xi \\ \xi \end{pmatrix}$

Where $\xi = \begin{pmatrix} \xi_1 \\ \xi_2 \end{pmatrix}$ refers to the spin with normalization $\xi^\dagger \xi = 1$.

Recall equation 7F.16 that the spin $\mathbf{S} = \frac{1}{2}\boldsymbol{\sigma}$, ($\hbar = 1$) with $\boldsymbol{\sigma}$ as the Pauli matrices. A particle with spin up along the z-axis would have $\xi = \begin{pmatrix} 1 \\ 0 \end{pmatrix}$, and spin down as $\xi = \begin{pmatrix} 0 \\ 1 \end{pmatrix}$.

A similar treatment for the anti-particle yields,

$$(10B.26) \quad (-\gamma_0 p^0 - m) \begin{pmatrix} v_L(p^0) \\ v_R(p^0) \end{pmatrix} = 0$$

$$\left[-\begin{pmatrix} 0 & I \\ I & 0 \end{pmatrix} m - m \begin{pmatrix} I & 0 \\ 0 & I \end{pmatrix} \right] \begin{pmatrix} v_L(p^0) \\ v_R(p^0) \end{pmatrix} = 0$$

$$-\begin{pmatrix} m & m \\ m & m \end{pmatrix} \begin{pmatrix} v_L(p^0) \\ v_R(p^0) \end{pmatrix} = 0$$

Solutions are of the type,

$$(10B.27) \quad v(p^0) \equiv \begin{pmatrix} v_L(p^0) \\ v_R(p^0) \end{pmatrix} = \sqrt{m} \begin{pmatrix} \eta \\ -\eta \end{pmatrix}$$

Where $\eta = \begin{pmatrix} \eta_1 \\ \eta_2 \end{pmatrix}$ refers to the spin with normalization $\eta^\dagger \eta = 1$. In this case, the spin up along the z-axis for an anti-particle is $\eta = \begin{pmatrix} 0 \\ 1 \end{pmatrix}$, for the spin down, $\eta = \begin{pmatrix} 1 \\ 0 \end{pmatrix}$. We see that in the Dirac equation, the spin of a Fermion (for both the particle and its anti-particle) emerges naturally.

Part 3

QUANTUM FIELD THEORY

Chapter 11

11A. The Essential QFT

An essential step in going from QM to QFT is to consider $\varphi(x)$ in the Klein-Gordon equation (10A.10a) as a scalar field operator. In other words, the K-G equation is an equation about operators rather than about a wave function. What we have done is a mapping between the field $\varphi(x)$ and the position vector x, an observable, hence an operator:

(11A.1) $x \rightarrow \varphi(x)$.

Pursuing our program of quantization, this step is often referred as second quantization. It is a bit misleading as in so far the fields haven't been quantized yet.

As it will turn out, the field will be the protagonist of our story, around which the plot (how to calculate stuff) will be easily understood. Nevertheless we retain the concept of a particle in a box of size L with,

(11A.2) $\psi(x) = \frac{1}{\sqrt{L}} e^{ipx}$

And eigenvalues,

(11A.3) $\hat{p}\psi(x) = -i\frac{\partial}{\partial x}\left(\frac{1}{\sqrt{L}}e^{ipx}\right) = p\left(\frac{1}{\sqrt{L}}e^{ipx}\right) = p\psi(x)$

Where we used equation 7C.7 for the momentum operator.

Boundary condition demands that

(11A.4) $e^{ipx} = e^{ip(x+L)}$

So $e^{ipL} = 1$ and $pL = 2\pi m$, where m is an integer. From this we can construct momentum states. For instance, for a two-particle state:

(11A.5) $\hat{p}\,|\,p_1 p_2> = (p_1 + p_2)\,|\,p_1 p_2>$

An arbitrary state $|\,\alpha>$ can be described in position coordinates as

(11A.6) $\psi_\alpha(x) = <x\,|\,\alpha>$

Or in momentum coordinates as,

(11A.7) $\tilde{\psi}_\alpha(p) = <p\,|\,\alpha>$

Now consider:

(11A.8) $\tilde{\psi}_\alpha(p) = <p\,|\,\alpha> = \int d^3x <p\,|\,x><x\,|\,\alpha>$

Where we used 7A.23.

Take the Fourier transform (Appendix J, equation JA.4) of equation 11A.8,

(11A.9) $\tilde{\psi}_\alpha(p) = \frac{1}{\sqrt{V}} \int d^3x\, e^{-ip\cdot x} \psi_\alpha(x)$

This implies,

(11A.10) $< p \mid x > = \frac{1}{\sqrt{V}}\, e^{-ip \cdot x}$

11B. Hamiltonian Formalism For Fields

To each observable, there is conjugate pair, and so we define the conjugate momentum as,

(11B.1) $\pi(x) = \partial\mathcal{L}/(\partial\varphi/\partial t)$

We define the Hamiltonian density as,

(11B.2a) $\mathcal{H}(\pi,\varphi) = \pi(x)(\partial\varphi(x)/\partial t) - \mathcal{L}(\varphi,\partial\varphi/\partial t,\nabla\varphi)$

Example:

Consider the Lagrangian density:

$\mathcal{L} = \frac{1}{2}(\partial\varphi/\partial t)^2 - \frac{1}{2}(\nabla\varphi)^2 - V(\varphi)$

Then the conjugate momentum density is:

(11B.2b) $\pi(x) = \partial\mathcal{L}/(\partial\varphi/\partial t) = \partial\varphi/\partial t \equiv \partial_t\varphi$

(11B.3) $\mathcal{H} = \frac{1}{2}\pi^2 + \frac{1}{2}(\nabla\varphi)^2 + V(\varphi)$

Where the Hamiltonian, $H = \int d^3x\, \mathcal{H}$.

11C. Canonical Quantization of the Fields

Recall in QM, canonical quantization tells us to take the coordinates q_a and the momenta p_a and promote them to operators ($\hbar = 1$).

(11C.1) $[q_a, p_b] = i\delta_{ab} \Rightarrow [\mathbf{q}, \mathbf{p}] = (2\pi)^3 \delta(\mathbf{q}\text{-}\mathbf{p})$ (in 3-D)

In field theory, we do the same for $\varphi(x)$ and $\pi(x)$.

(11C.2a) $[\varphi_a(x), \pi_b(y)] = [\varphi_a(x), \partial_t \varphi_b(y)] = i\delta_{ab}\, \delta^{(3)}(x\text{-}y)$

Where we've taken from equation 10A.14, which is now a field operator:

$\varphi_a(x) = \int \frac{d^3k}{(2\pi)^{3/2}(2E_p)^{1/2}} (a^\dagger{}_p e^{ipx} + a_p e^{-ipx})$

As was mentioned above the simplest free field theory is the Klein-Gordon equation (K-G) for a real scalar, where m has dimension of mass.

(11C.3) $\partial_\mu \partial_\mu \varphi + m^2 \varphi = 0$

Solving for φ, we do a Fourier transformation,

(11C.4) $\varphi(x,t) = \int d^3 p (2\pi)^{-3} e^{\,ip \cdot x} \varphi(p,t)$

Substituting (11C.4) into the K-G equation,

(11C.5) $[\partial^2/\partial t^2 + (p^2 + m^2)] \varphi(p,t) = 0$

This is the equation for a harmonic oscillator (see equation 7D.3)with frequency,

(11C.6) $\omega_p = (p^2 + m^2)^{\frac{1}{2}}$

The solutions to the classical K-G equations are superpositions of simple harmonic oscillators (SHO).

11D. Free Field Theory

From the Klein-Gordon equation, recall the following,

(11D.1) $H = \frac{1}{2}\int d^3x[\pi^2 + (\nabla\varphi)^2 + m^2\varphi^2]$, (equation 11B.3)

And equations 11C.6, 11C.2,

(11D.2) $\omega_p = + (\mathbf{p}^2 + m^2)^{\frac{1}{2}}$ and $[\varphi(\mathbf{x}), \pi(\mathbf{y})] = i\delta^{(3)}(\mathbf{x}-\mathbf{y})$

We take the definition of the creation operators $a_p{}^\dagger$ and annihilation operators a_p, and use that to define a scalar field and scalar particle, spin zero as,

(11D.3) $\varphi(x) = \int d^3p(2\pi)^{-3}(2\omega_p)^{-\frac{1}{2}} [a_p e^{ip\cdot x} + a_p{}^\dagger e^{-ip\cdot x}]$

And its conjugate field (equation 11B.2b),

(11D.4) $\pi(x) = \int d^3p(2\pi)^{-3}(-i)(\omega_p/2)^{\frac{1}{2}}[a_p e^{ip\cdot x} - a_p{}^\dagger e^{-ip\cdot x}]$

Note that both $\varphi(x)$ and $\pi(x)$ are hermitian operators, with $\varphi^\dagger = \varphi$ and $\pi^\dagger = \pi$. Recall the commutator, from equation 11C.1, repeated below,

(11D.5) $[a_p, a_q{}^\dagger] = (2\pi)^3\delta^{(3)}(\mathbf{p}-\mathbf{q})$,

Note: for $\mathbf{p} = \mathbf{q}$, $[a_p, a_p{}^\dagger] = (2\pi)^3\delta^{(3)}(0)$

Now substitute equations 11D.3 and 11D.4 into 11D.1,

(11D.6)
$H = \frac{1}{2}\int d^3x [\pi^2 + (\nabla\varphi)^2 + m^2\varphi^2]$

$= \frac{1}{2}\int d^3x\, d^3p\, d^3q(2\pi)^{-6} \{-(\omega_q\omega_p/2)^{-\frac{1}{2}}$

$$X\ [a_pe^{ip\cdot x} - a_p{}^\dagger e^{-ip\cdot x}][a_qe^{iq\cdot x} - a_q{}^\dagger e^{-iq\cdot x}]$$
$$+i2^{-1}(\omega_p\omega_q)^{-\frac{1}{2}}[\mathbf{p}a_pe^{ip\cdot x} - \mathbf{p}a_p{}^\dagger e^{-ip\cdot x}][\mathbf{q}a_qe^{iq\cdot x} - \mathbf{q}a_q{}^\dagger e^{-iq\cdot x}]$$
$$+m^22^{-1}(\omega_p\omega_q)^{-\frac{1}{2}}[a_pe^{ip\cdot x} + a_p{}^\dagger e^{-ip\cdot x}][a_qe^{iq\cdot x} + a_q{}^\dagger e^{-iq\cdot x}]\}$$

Using the definition of the delta function (appendix J, equation JB.5),

(11D.7) $\delta(\mathbf{x} - \mathbf{y}) = (2\pi)^{-3}\int d^3k e^{ik(x-y)}$

We integrate over x and q.

(11D.8)

$$H = (\tfrac{1}{4})\int d^3p(2\pi)^{-3}(\omega_p)^{-1}\ [(-\omega_p{}^2 + p^2 + m^2)(a_pa_{-p} + a_p{}^\dagger a_{-p}{}^\dagger)$$
$$+(\omega_p{}^2 + p^2 + m^2)(a_pa_p{}^\dagger + a_p{}^\dagger a_p)]$$

But $\omega_p{}^2 = p^2 + m^2$, therefore the first term in the square bracket vanishes, and we are left with,

(11D.9) $H = \tfrac{1}{2}\int d^3p\ (2\pi)^{-3}\omega_p(a_pa_p{}^\dagger + a_p{}^\dagger a_p)$

This is the energy of an infinite number of uncoupled SHO, as expected. Also from the commutation relationship (11D.5) we get,

(11D.10) $H = \int d^3p[\ (2\pi)^{-3}\omega_p(a_pa_p{}^\dagger + \tfrac{1}{2}(2\pi)^3\delta^{(3)}(0)\ ]$

The last term is problematic as it is infinite.

11E. The Vacuum

Define the vacuum as in the case of the SHO, (equation 7D.30),

(11E.1) $a_p|0> = 0$, for all **p**.

Its energy is then,

(11E.2) $H|0> = E_0|0> = [\int d^3p\,\omega_p\delta^{(3)}(0)]|0>$

Note that we have infinity coming from two sources:

(i) from the delta function, because space extends to infinity in all directions - these are called infra-red (IF) divergences.

(ii) Consider a box of size L. In the limit that $L \to \infty$,

(11E.3) $(2\pi)^3\delta^{(3)}(0) = \lim \int_{-L/2}^{+L/2} d^3x\,e^{ip.x} = V$

Where V is the volume of the box.

(11E.4) Therefore, $E_0 = \frac{1}{2}V \int d^3p\,(2\pi)^{-3}\omega_p$

This indicates that we should work with energy density,

(11E.5) $E = E_0/V = \frac{1}{2} \int d^3p\,(2\pi)^{-3}\omega_p$

This energy density is still infinite due to space being infinitesimally small (as $p \to \infty$, x gets smaller and smaller, that is, we have bigger and bigger oscillations over smaller and smaller distances. This is the second source of infinity, which is called the ultra-violet (UV) divergence. Here we have assumed that our theory is valid over arbitrarily short distance scales.

Since we are only interested in energy differences, we remove the infinities and redefine,

(11E.6) $H = \int d^3p \, (2\pi)^{-3} \omega_p a_p a^\dagger_p$

This yields,

(11E.7) $H|0> = 0$

11F. Relativistic Normalization

(11F.1) Recall that the vacuum is normalized, $<0|0> = 1$

(11F.2) And for the 1-particle state, we have,

$|p> = a_p^\dagger|0>$ and $<p|q> = (2\pi)^3 \delta^{(3)}(p-q)$

The problem here is that it isn't clear if this expression is Lorentz invariant.

Claim 1:

(11F.3) $\int \frac{d^3p}{2E_p}$, where $E_p = \sqrt{p^2 + m^2}$ is Lorentz invariant.

Proof:

(i) $\int d^4p$ is Lorentz invariant.

(ii) $p_0^2 = E_p^2 = p^2 + m^2$ is also Lorentz invariant.

(iii) Combining (i) and (ii) is Lorentz invariant. That is, consider the integral,

$\rightarrow \int d^4p \, \delta^{(3)}(p_0^2 - \mathbf{p}^2 - m^2)$

Where $E_p \equiv p_0$. Using the identity, (appendix J, equation JB.4))

Define $f(p) \equiv p_0{}^2 - \mathbf{p}^2 - m^2 \rightarrow f'(p) = 2p_0$

Integrate over p_0 and substitute (ii),

$\rightarrow \int \frac{d^3p}{2E_p}$ is Lorentz invariant.

Claim 2:

(11F.4) $2E_p\, \delta^{(3)}(\mathbf{p-q})$ is Lorentz invariant.

Proof:

(11F.5) $\int \frac{d^3p}{2E_p} 2E_p\, \delta^{(3)}(\mathbf{p-q}) = 1$ (equation JB.1)

Since 1 is invariant, $\delta^{(3)}(\mathbf{p-q})$ is also Lorentz invariant.

QED

This motivates us to normalize our states as

(11F.6a) $|p> \rightarrow \sqrt{2E_p}\, |p>$

(11F.6b) Then $<p|q> = (2\pi)^3\, 2E_p\, \delta^{(3)}(\mathbf{p-q})$, which is now Lorentz invariant.

This is the justification for the normalized factor in equation 10A.14.

11G. The Electrodynamics of a Charged Scalar Field

The corresponding equation for a charged scalar particle is obtained from the K-G equations by making the substitution,

(11G.1) $i\partial_\mu \rightarrow i\partial_\mu - qA_\mu$

Where q is the charge, and A_μ is the electromagnetic field (equation 3G.1).

Substitute equation 11G.1 into the K-G equation (11C.3).

(11G.2)

$$(\partial^2 + m^2)\phi = 0$$

$$\rightarrow \{(i\partial_\mu - qA_\mu)(i\partial_\mu - qA_\mu) - m^2\}\phi = 0$$

The presence of the complex number dictates that a solution of equation 11G.2 is necessarily complex. Thus a charged particle of zero spin in an electromagnetic field can be described by a complex wave function,

(11G.3a) $\phi = \frac{1}{\sqrt{2}}(\phi_1 + i\phi_2)$

(11G.3b) $\phi^* = \frac{1}{\sqrt{2}}(\phi_1 - i\phi_2)$

Note that since zero spin particles are bosons, the fields ϕ and ϕ^* commute. Also if ϕ is a solution for a particle carrying a charge of –q, then ϕ^* is for a particle with charge +q.

Chapter 12

Propagators and Feynman Diagrams

Recall that in QM, the wave function, which satisfies the Schrödinger equation (7C.9), plays a fundamental role in calculating probability amplitudes under the Born rule (7B.5). But in QFT the Schrödinger equation is replaced by the Klein-Gordon equation in order for the theory to be compatible with SR, and the wave function is now a field operator (section 11C). So how are we going to calculate probability amplitudes? As this story will unfold, propagators will come to our rescue. Moreover, propagators will give us more than what we ask for.

12A. Time-Evolution Operator

Our first step is to define the time evolution operator,

(12A.1) $| \psi(t_2) > = U(t_2, t_1) | \psi(t_1) >$

So the operator U is taken as evolving the state ψ from time t_1 to time t_2. Here we are not moving a particle through time. In reality it is the quantum state that is evolving in time. The hope is that it will also represent the motion of a particle.

Properties of this time-evolution operator are:

(12A.2) $U(t_1, t_1) = 1$

When nothing happens, the operator U is unity.

(12A.3) $U(t_3, t_2) U(t_2, t_1) = U(t_3, t_1)$

This is the composition law. A time evolution can be broken up into smaller steps.

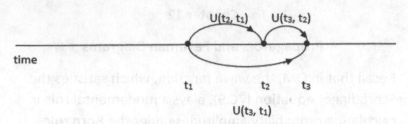

Fig. 12A.1

(12A.4) The time-evolution operator itself obeys the Schrödinger equation (7C.9),

$$i \frac{d}{dt_2} U(t_2, t_1) = H\, U(t_2, t_1)$$

Proof:

Differentiate equation 12A.1 with respect to t_2 and multiply both sides by i :

$$i \frac{d}{dt_2} |\psi(t_2)> = i\left(\frac{d}{dt_2} U(t_2, t_1)\right) |\psi(t_1)>$$

From equation 7C.9, with $\hbar = 1$,

$$\rightarrow i \frac{d}{dt_2} |\psi(t_2)> = H |\psi(t_2)>$$

$$= H\, U(t_2, t_1)| \psi(t_1)> \text{ (using 12A.1)}$$

Equating the two equations:

$$\rightarrow i\left(\frac{d}{dt_2} U(t_2, t_1)\right) |\psi(t_1)> = H\, U(t_2, t_1)| \psi(t_1)>$$

Therefore,

$$i \frac{d}{dt_2} U(t_2, t_1) = H\, U(t_2, t_1)$$

QED

Note that

(12A.5) $U(t_2, t_1)U(t_1, t_2) = 1$

This means that if we take a state from $\psi(t_1)$ to a state $\psi(t_2)$, we should be able to reverse by an inverse operation. This means $U(t_2, t_1) = U^{-1}(t_1, t_2)$ (Fig. 12A.2).

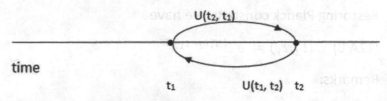

time

t_1 $U(t_1, t_2)$ t_2

Fig. 12A.2

As a consequence, we get that

(12A.6) $U^\dagger(t_2, t_1)U(t_2, t_1) = 1$,

that is, the time-evolution operator is unitary.

Proof:

Differentiate equation 12A.6 with respect to t_2:

$$\rightarrow \frac{d}{dt_2}\{U^\dagger(t_2, t_1)U(t_2, t_1)\} = \frac{dU^\dagger}{dt_2}U + U^\dagger \frac{dU}{dt_2}$$

Using (12A.4),

$$\rightarrow \frac{dU}{dt_2} = \frac{HU}{i} \text{ and } \frac{dU^\dagger}{dt_2} = -\frac{U^\dagger H}{i}$$

The second part we used $(HU)^\dagger = U^\dagger H^\dagger = U^\dagger H$

We get,

$$\rightarrow \frac{d}{dt_2}\{U^\dagger(t_2,t_1)U(t_2,t_1)\} = -\frac{U^\dagger HU}{i} + \frac{U^\dagger HU}{i} = 0$$

QED.

Also we have $U^\dagger(t_2,t_1) = U^{-1}(t_1,t_2)$

Solving equation (12A.4), we get

(12A.7) $U(t_2,t_1) = e^{-iH(t_2-t_1)}$

Restoring Planck constant, we have

(12A.8) $U(t_2,t_1) = e^{-iH(t_2-t_1)/\hbar}$

Remarks:

a) Even though this wasn't mentioned, we are working in what is called the Schrödinger picture. The above equation 7C.9 should read:

(12A.9) $i\hbar \frac{d|\psi_S(t)>}{dt} = H_S |\psi_S(t)>$

Where the subscript "S" stands for "Schrödinger."

Note that the features in this picture are: (i) quantum states are time-dependent; (ii) and the operators are time-independent.

b) The argument in the exponential of equation 12A.8 can be written as $-iH_S(t_2 - t_1) \rightarrow -iH_S \int_{t_1}^{t_2} dt$. This is allowed as the Hamiltonian H_S in the Schrödinger pictures is time-independent and can be taken out of the integral. As we shall see next, in the Heisenberg picture, this will no longer apply. And handling the Hamiltonian will require extra care.

c) We are dealing so far with free particles. So the potential V = 0, and therefore we have L = T, and H = T. That is, the Lagrangian L = H, the Hamiltonian. Equation 12A.8 can then be written as

(12A.10) $U(t_2, t_1) = e^{-iS/\hbar}$

Where $S = \int_{t_1}^{t_2} L\, dt$, is the action (equation 2A.4).

d) Equation 12A.1 can now be written as

(12A.11) $|\psi(t_2)\rangle = e^{-iH(t_2 - t_1)/\hbar}|\psi(t_1)\rangle$.

12B. The Heisenberg Picture

A situation that is more analogous to Classical physics is the Heisenberg picture in which the quantum states are time-independent and the observables, time-dependent. Another reason is that once we move on to the interaction picture we will deal with a Fock space in which the quantum states are defined in terms of particle number. Likewise as in the Heisenberg picture, these states are time-independent. So how do we define the Heisenberg picture?

Our concern will be with the state $\psi_S(0)$, after all if we can measure the state at time t = 0, we can always get $\psi_S(t)$ with the time-evolution operator.

(12B.1) $\psi_S(t) = U(t, 0)\psi_S(0)$ from equation 12A.1

$\qquad = e^{-iH(t)}\psi_S(0)$ from equation 12A.11

In the last line we used $\hbar = 1$. Now in the Heisenberg picture the operators are time dependent. For any operator $O_H(t)$, we want the expectation value to be invariant in both pictures, that is,

(12B.2) $< \psi_H \,|\, O_H(t) \,|\, \psi_H > = < \psi_S(t) \,|\, O_S \,|\, \psi_S(t) >$

In the RHS, we can substitute (from equation 12A.1),

(12B.3) RHS $= < \psi_S(t) \,|\, O_S \,|\, \psi_S(t) >$

$$= < \psi_S(0) \,|\, U^\dagger(t,0)\, O_S\, U(t,0) \,|\, \psi_S(0) >$$

The quantum state are independent of time in the Heisenberg picture. So we can equate $\psi_H \equiv \psi_S(0)$, that is, we freeze the states at time t = 0, then we can identify with the LHS,

(12B.4) $O_H(t) = U^\dagger(t,0)\, O_S\, U(t,0)$

From this we get the Heisenberg equation of motion (the equivalent of the Schrödinger equation). Differentiate the above equation:

(12B.5) $\dfrac{dO_H(t)}{dt} = \dfrac{dU^\dagger}{dt} O_S U + U^\dagger O_S \dfrac{dU}{dt}$

Using equation 12A.4,

(12B.6) $\dfrac{dU}{dt} = \dfrac{HU}{i}$ and $\dfrac{dU^\dagger}{dt} = -\dfrac{U^\dagger H}{i}$

Then,

(12B.7) $\dfrac{dO_H(t)}{dt} = \dfrac{1}{i}(-U^\dagger H\, O_S\, U + U^\dagger O_S\, HU)$

Assuming that H commutes with U and $U^\dagger$,

(12B.8) $\frac{dO_H(t)}{dt} = \frac{1}{i}(-H\,U^\dagger O_S U + U^\dagger O_S U\,H)$

Using the above result (12B.4) for operators and 7A.13a,

(12B.9) $\frac{dO_H(t)}{dt} = (-HO_H + O_H H) = \frac{1}{i}[O_H(t), H]$

This is the equation of motion in the Heisenberg picture. Note that the Hamiltonian is the same in both pictures:

(12B.10) $H_H(t) = U^\dagger(t,0)\,H_S\,U(t,0)$ (equation 12B.4)

$\qquad = H_S\,U^\dagger(t,0)\,U(t,0)$ (H commutes with $U^\dagger$ and U)

$\qquad = H_S$ (equation 12A.5)

12C. Green Functions

Green functions stem from inhomogeneous linear equations.

(12C.1) $\hat{L}\,x(t) = f(t)$

Where $\hat{L}$ is a linear operator defined by,

(12C.2) $\hat{L}G(t,u) = \delta(t-u)$

And $G(t,u)$ is the Green function.

The strategy is that equation 12C.1 is often very difficult to solve due to what kind of function f(t) could be. Equation 12C.2 contains instead a delta function, which makes it easier to solve. So the question is: once we solve 12C.2 for the Green function, how does that make the solving of equation 12C.1 easier? We shall demonstrate that with an example, the case of the harmonic oscillator.

Suppose we have an oscillator under the influence of a time-dependent force $f(t)$. The equation is then,

(12C.3) $m\frac{d^2}{dt^2}x(t) + kx(t) = f(t)$

In this case, the linear operator is $\hat{L} = m\frac{d^2}{dt^2} + k$

We are now looking for a Green function that will satisfy (12C.2):

(12C.4) $\hat{L}G(t,u) = (m\frac{d^2}{dt^2} + k)G(t,u) = \delta(t-u)$

Using the definition of the delta function (appendix J, equation JB.1) we can write the time-dependent force $f(t)$ as,

(12C.5) $f(t) = \int_0^\infty du f(u)\delta(t-u)$

The solution is readily obtained,

(12C.6) $x(t) = \int_0^\infty du G(t-u)f(u)$

Proof:

Multiply equation 12C.6 by the linear operator $\hat{L}$:

$\hat{L}x(t) = \int_0^\infty du \hat{L}G(t-u)f(u)$

$= \int_0^\infty du\,\delta(t-u)f(u)$ (equation 12C.4)

$= f(t)$ (equation 12C.5)

QED

Note: Whatever Green function is a solution to equation 12C.5 for $f(t)$, we readily have the solution to equation 12C.3 for $x(t)$.

12D. Propagators

Our next step is to link the Green's functions to Quantum Field Theory. We know that the main object in QFT is the field operator $\phi(x)$ (from equation 11C.2) in which case $[\phi_a(x), \pi_b(y)] = i\delta_{ab}\, \delta^{(3)}(x-y)$). So we propose the following definition.

$$(12D.1)\ \phi(x, t_x) = \int dy\ G^+(x, t_x, y, t_y)\phi(y, t_y)$$

We will later on show that $G^+(x, t_x, y, t_y)$ is a Green's function. But before that, we can think of it as,

$$(12D.2)\ G^+ = G,\ t_x > t_y$$

$$= 0,\ t_x < t_y$$

In other words, t_x occurs later than t_y. (Note: $G^- = G$, $t_y > t_x$, for t_x occurring earlier than t_y). We can also write the above as,

$$(12D.3)\ G^+ = \theta(t_x - t_y)G$$

Where $\theta(t_x - t_y)$ is the Heavyside function (appendix J, equation JC.1). We can now reinterpret $\phi(y, t_y)$ as the amplitude to find a particle at (y, t_y). Similarly, we define $\phi(x, t_x)$ as the amplitude to find a particle at a later position (x, t_x). From this then the propagator $G^+(x, t_x, y, t_y)$ is the probability amplitude that a particle in a quantum state starts at (y, t_y) and *propagates* to point (x, t_x). Hence we will postulate that the green function can be written as,

(12D.4) $G^+(x, t_x, y, t_y) = \theta(t_x - t_y) < x(t_x)|y(t_y) >$

Note: in QM, we have $\phi(x, t_x) = < x|\phi(t) >$ is the amplitude of a particle found at (x, t_x) regardless of where it started. Therefore, the propagator contains **more** information since it also tells us where it started.

So far we assume that the propagator $G^+(x, t_x, y, t_y)$ is a Green's function of the Schrödinger equation. We will now demonstrate that.

Starting with equations (12A.1) and (12D.4),

(12D.5)

$$G^+(x, t_x, y, t_y) = \theta(t_x - t_y) < x|U(t_x - t_y)|y >$$
$$= \theta(t_x - t_y) < x|e^{-iH(t_x - t_y)}|y >$$

Where we used in the last line equation 12A.7. We can expand this expression in terms of the energy eigenstates, that is, $H|n> = E_n|n>$,

(12D.6)

$$G^+(x, t_x, y, t_y)$$
$$= \theta(t_x - t_y) \sum_n < x|e^{-iH(t_x - t_y)}|n >< n|y >$$
$$= \theta(t_x - t_y) \sum_n e^{-iE_n(t_x - t_y)} < x|n >< n|y >$$
$$= \theta(t_x - t_y) \sum_n e^{-iE_n(t_x - t_y)} \phi_n(x) \phi_n^*(y)$$

Where $\phi_n(x) = <x|n>$ and $\phi_n^*(y) = <n|y>$

Our next step is to show that the amplitude in 12D.4,

$<x(t_x)|y(t_y)>$, is truly the Green's function of the Schrödinger equation.

Proof:

Recall for the Green's function it has to satisfy equation 12C.1, and in this case the linear operator in the Schrödinger (equation7C.9) is,

$$(12D.7) \ L = H_x - i\frac{\partial}{\partial t_x}$$

Where the linear operator deals only with the x-coordinates – the y-coordinates are fixed as dummy variables.

Start with the first term in 12D.7 acting on G^+,

(12D.8)

$$H_x \ G^+(x, t_x, y, t_y)$$

$$= H_x \ \theta(t_x - t_y) < x(t_x)|y(t_y)> \ \text{(equation 12D.4)}$$

$$= \theta(t_x - t_y) \ \sum_n E_n e^{-iE_n(t_x - t_y)} \ \phi_n(x) \ \phi_n^*(y)$$

The last line was obtained from equation 12D.6

For the second term in 12D.7,

$$(12D.9) \ i\frac{\partial}{\partial t_x} \ G^+(x, t_x, y, t_y)$$

$$= i\frac{\partial}{\partial t_x} \ \{\theta(t_x - t_y) \ \sum_n e^{-iE_n(t_x - t_y)} \ \phi_n(x) \ \phi_n^*(y)\}$$

From Appendix J, equation JC.2, we have,

(12D.10) $\frac{\partial}{\partial t_x} \theta(t_x - t_y) = \delta(t_x - t_y)$

So carrying the differentiation of (12D.9), we get

(12D.11) $i \frac{\partial}{\partial t_x} G^+(x, t_x, y, t_y)$

$= i\delta(t_x - t_y) \sum_n e^{-iE_n(t_x - t_y)} \phi_n(x) \phi_n^*(y)$

$\quad + i\{ \theta(t_x - t_y) \sum_n (-iE_n) e^{-iE_n(t_x - t_y)} \phi_n(x) \phi_n^*(y)$

To evaluate equation (12D.7), we subtract equation (12D.8) and (12D.11),

(12D.12) $\left(H_x - i \frac{\partial}{\partial t_x} \right) G^+(x, t_x, y, t_y)$

$= -i\delta(t_x - t_y) \sum_n e^{-iE_n(t_x - t_y)} \phi_n(x) \phi_n^*(y)$

Working backward from 12D.6 to 12D.4,

(12D.13) $\sum_n e^{-iE_n(t_x - t_y)} \phi_n(x) \phi_n^*(y)$

$= < x(t_x) | y(t_y) >$

$= \int dp \, e^{ip \cdot x} e^{-ip \cdot y}$ (equation 11A.10)

$= \int dp \, e^{ip \cdot (x-y)} = \delta(x - y)$

Where in the last line we used equation JB.1 in appendix J.

Substituting that into (12D.12), we get,

(12D.14) $\left(H_x - i \frac{\partial}{\partial t_x} \right) G^+(x, t_x, y, t_y)$

$= -i\delta(t_x - t_y)\delta(x - y)$

Which is of the form of equation (12C.2)

QED

12E. Feynman Propagators

We are now in the position of defining the Feynman propagator. We must keep in mind that we are dealing with two crucial facts observed in nature: a) particles can be annihilated and created; b) creation/annihilation events deal with particles and anti-particles. In other words, we start with a system in its ground state, denoted by $|\Omega>$. The propagator will be made of two parts: the first part applies on a particle created at y that propagates to x where it is destroyed; a second part for an anti-particle created at x and propagates to y where it is destroyed.

(12E.1)

$$G^+(x,\ y) = \theta(x^0 - y^0) < \Omega|\ \phi_n(x)\ \phi_n^\dagger(y)\ |\Omega>$$
$$+ \theta(y^0 - x^0) < \Omega|\phi_n^\dagger(y)\ \phi_n(x)\ |\Omega>$$

To deal with such a case, we will use a time-ordering symbol T, defined as,

(12E.2a) $G^+(x,\ y) \equiv < \Omega|\ T\phi_n(x)\ \phi_n^\dagger(y)\ |\Omega>$

Where x^0 is later, which means that if $x^0 > y^0$, we have

(12E.2b) $T\phi_n(x^0)\ \phi_n(y^0) = \phi_n(x^0)\ \phi_n(y^0)$

For y^0 later, if $y^0 > x^0$ then

(12E.2c) $T\phi_n(x^0)\ \phi_n(y^0) = \phi_n(y^0)\ \phi_n(x^0)$

This insures that the **later** operators are always on the **left**.

Our next step is to define the free Feynman propagator as,

(12E.3) $\Delta(x, y) \equiv\, < 0|\, T\phi_n(x)\, \phi_n^\dagger(y)\, |0 >$

Which acts on the vacuum ($|0 >$) instead of the ground state $|\Omega >$. This is a simpler and easier case to calculate. Let us work out an expression of the free Feynman propagator for a scalar field.

<u>Step 1</u> : Consider the field $\phi_n^\dagger(y)$ operating on the vacuum $|0 >$.

(12E.4) $\phi_n^\dagger(y)\, |0 >$

$$= \int \frac{d^3p}{(2\pi)^{\frac{3}{2}}(2E_p)^{\frac{1}{2}}} \left(a_p^\dagger |0 > e^{ipy} + \; b_p |0 > e^{-ipy} \right.$$

Where we used equation 10A.14.

Since $b_p|0 > = 0$, the second term vanishes. We are left with,

(12E.5) $\phi_n^\dagger(y)\, |0 > = \int \frac{d^3p}{(2\pi)^{\frac{3}{2}}(2E_p)^{\frac{1}{2}}} a_p^\dagger |0 > e^{ipy}$

$$= \int \frac{d^3p}{(2\pi)^{\frac{3}{2}}(2E_p)^{\frac{1}{2}}} |p > e^{ipy}$$

To get the other half of equation 12E.1, take the complex conjugate of equation 12E.5 and swap $y \rightarrow x$ and $p \rightarrow q$.

(12E.6) $< 0|\, \phi_n(x) = \int \frac{d^3q}{(2\pi)^{\frac{3}{2}}(2E_q)^{\frac{1}{2}}} < q|\, e^{-iqx}$

Now sandwich together equations 12E.5 and 12E.6,

(12E.7) $G_0^+(x, y) \equiv\, < 0|\, \phi_n(x)\, \phi_n^\dagger(y)\, |0 >$

$$= \int \frac{d^3p \, d^3q}{(2\pi)^3 (2E_p 2E_q)^{\frac{1}{2}}} e^{-iqx + ipy} < q \, | \, p >$$

$$= \int \frac{d^3p \, d^3q}{(2\pi)^3 (2E_p 2E_q)^{\frac{1}{2}}} e^{-iqx + ipy} \delta^{(3)}(q - p)$$

Where in the last line we used equation 7A.23b. Doing the q integral, we get

(12E.8) $G_0^+(x, y) = < 0 \, | \, \phi_n(x) \, \phi_n^\dagger(y) \, | \, 0 >$

$$= \int \frac{d^3p}{(2\pi)^3 (2E_p)} e^{-ip(x - y)}$$

This corresponds to a particle being created at at point (y^0, y) and propagating to point (x^0, x), where it is annihilated.

We need the reverse process.

Step 2 :

(12E.9) $G_0^-(x, y) = < 0 \, | \, \phi_n^\dagger(y) \phi_n(x) \, | \, 0 >$

We proceed as before:

(12E.10) $\phi_n(x) \, | \, 0 >$

$$= \int \frac{d^3p}{(2\pi)^{\frac{3}{2}} (2E_p)^{\frac{1}{2}}} (a_p \, | \, 0 > e^{-ipx} + b_p^\dagger \, | \, 0 > e^{ipx}$$

Since $a_p \, | \, 0 > = 0$, the first term vanishes. We have,

(12E.11) $\phi_n(x) \, | \, 0 > = \int \frac{d^3p}{(2\pi)^{\frac{3}{2}} (2E_p)^{\frac{1}{2}}} b_p^\dagger \, | \, 0 > e^{ipx}$

$$= \int \frac{d^3p}{(2\pi)^{\frac{3}{2}} (2E_p)^{\frac{1}{2}}} \, | \, p > e^{ipx}$$

Again we take the complex conjugate, swapping $x \to y$ and $p \to q$.

(12.E12) $< 0 | \phi_n^\dagger(y) = \int \frac{d^3q}{(2\pi)^{\frac{3}{2}}(2E_q)^{\frac{1}{2}}} < q | e^{-iqy}$

Sandwiching the two and doing the integral over q, we get,

(12E.12) $G_0^-(x, y) = < 0 | \phi_n^\dagger(y)\phi_n(x) | 0 >$

$$= \int \frac{d^3p}{(2\pi)^3(2E_p)} e^{ip(x-y)}$$

This corresponds to an anti-particle created at point $(x^0, \boldsymbol{x})$, propagating to point $(y^0, \boldsymbol{y})$.

Putting together, we get the final result for the free Feynman propagator (12E.3):

(12E.13) $\Delta(x - y) \equiv < 0 | T\phi_n(x) \phi_n^\dagger(y) | 0 >$

$$= \int \frac{d^3p}{(2\pi)^3(2E_p)} \{\theta(x^0 - y^0)e^{-ip(x-y)}$$

$$+ \theta(y^0 - x^0)e^{ip(x-y)}\}$$

Since the Feynman propagator depends on the difference of the two points, we have written $\Delta(x, y) \to \Delta(x - y)$.

So far so good but is it a Green function? Does it satisfy equation 12C.2? We will now demonstrate that it satisfies the following equation,

(12E.14) $(\partial^2 + m^2)\Delta(x - y) = -i\delta^{(4)}(x - y)$

Where the linear operator $L = \partial^2 + m^2$ is taken from the Klein-Gordon equation (equation 10A.10a) instead of the

Schrödinger equation. This is in line with going from QM to QFT.

Proof:

We will have a change of variables:

$$x \rightarrow x, \ x^0 \rightarrow t; \ y \rightarrow x' \ y^0 \rightarrow t'$$

Also, we will consider the system in its ground state $|\Omega>$. We start with evaluating:

(12E.15) $(\partial^2 + m^2) < \Omega | T\{\phi_n(x)\,\phi_n(x')\} | \Omega >$

Recall that

(12E.16) $\partial^2 \equiv \partial_t^2 - \nabla^2$ (Equation EA.16, appendix E)

First we will work out the time derivatives of equation 12E.15, that is,

(12E.17) $\partial_t < \Omega | T\{\phi_n(x)\,\phi_n(x')\} | \Omega >$

$= \partial_t \{\theta(t - t') < \Omega | \phi_n(x)\,\phi_n(x') | \Omega >$

$\quad + \theta(t' - t) < \Omega | \phi_n(x')\,\phi_n(x) | \Omega >\}$

Note that the derivative is taken with respect to t, and not t'.

$= \partial_t \{\theta(t - t')\} < \Omega | \phi_n(x)\,\phi_n(x') | \Omega >$

$\quad\quad + \theta(t - t') \{< \Omega | \partial_t \phi_n(x)\,\phi_n(x') | \Omega >\}$

$+ \partial_t \{\theta(t' - t)\} < \Omega | \phi_n(x')\,\phi_n(x) | \Omega >\}$

$\quad\quad + \theta(t' - t)\{< \Omega | \phi_n(x')\,\partial_t\,\phi_n(x) | \Omega >\}$

From Appendix J, equation JC.2, we have,

$$\partial_t\{\theta(t - t')\} = \delta(t - t')$$

$$\text{and } \partial_t\{\theta(t' - t)\} = -\delta(t - t')$$

The 1st and 3rd terms form a commutation relation. Also, we can combine the 2nd and 4th term using the definition of the time-ordering operator. Putting altogether, we get,

(12E.18) $\partial_t < \Omega | T\{ \phi_n(x) \phi_n(x')\} | \Omega >$

$$= < \Omega | T\{ \partial_t\phi_n(x) \phi_n(x')\} | \Omega >$$

$$+ \delta(t - t') < \Omega | [\phi_n(x'), \phi_n(x)] | \Omega >$$

Taking again the second time derivative will yield a similar calculation:

(12E.19) $\partial_t^2 < \Omega | T\{ \phi_n(x) \phi_n(x')\} | \Omega >$

$$= < \Omega | T\{ \partial_t^2\phi_n(x) \phi_n(x')\} | \Omega >$$

$$+ \delta(t - t') < \Omega | [\partial_t\phi_n(x), \phi_n(x')] | \Omega >$$

However from equation 11C.2,

(12E.20) $[\partial_t\phi_n(x), \phi_n(x')] = -i\,\delta^{(3)}(x - x')$

Substituting,

(12E.21) $\partial_t^2 < \Omega | T\{ \phi_n(x) \phi_n(x')\} | \Omega >$

$$= < \Omega | T\{ \partial_t^2\phi_n(x) \phi_n(x')\} | \Omega > -i\,\delta^{(4)}(x - x')$$

The expression 12E.15 now reads as,

(12E.22) $(\partial^2 + m^2) < \Omega | T\{ \phi_n(x) \phi_n(x')\} | \Omega >$

$$= < \Omega | T\{ (\partial^2 + m^2) \phi_n(x) \phi_n(x')\} | \Omega > -i\,\delta^{(4)}(x - x')$$

In the free field picture, we have

(12E.23) $(\partial^2 + m^2) \phi(x) = 0$ (K-G equation10A.10a)

Using equation 12E.3, equation 12E.23 becomes,

$(\partial^2 + m^2) \Delta(x - y) = -i\delta^{(4)}(x - y)$

QED

Now that we have established that the free Feynman propagator is a Green function of the K-G equation, our next step is to put equation 12E.13, repeated below, in a more suitable form for later calculations.

(12E.24) $\Delta(x - y) = < 0 | T\phi_n(x) \phi_n^\dagger(y) | 0 >$

$= \int \frac{d^3p}{(2\pi)^3(2E_p)} \{\theta(\tau)e^{-ip(x-y)} + \theta(-\tau)e^{ip(x-y)}\}$

Where we use $x^0 - y^0 = \tau$

(12E.25a) Expand, $p(x - y) = p^0\tau - \vec{p} \cdot (\vec{x} - \vec{y})$
$= E_p\tau - \vec{p} \cdot (\vec{x} - \vec{y})$

(12E.25b) Where $E_p{}^2 = (|\vec{p}|)^2 + m^2$ (Equation 5G.26)

Substitute 12E.25 into 12E.24, we get

(12E.26) $\Delta(x - y)$

$= \int \frac{d^3p}{(2\pi)^3(2E_p)} \{\theta(\tau)e^{-ip(x-y)} + \theta(-\tau)e^{ip(x-y)}\}$

$= \int \frac{d^3p}{(2\pi)^3(2E_p)} \{\theta(\tau)e^{-i(E_p\tau - \vec{p}\cdot(\vec{x}-\vec{y}))}$
$+ \theta(-\tau)e^{i(E_p\tau - \vec{p}\cdot(\vec{x}-\vec{y}))}\}$

$$= \int \frac{d^3p}{(2\pi)^3(2E_p)} \{ e^{i\vec{p}\cdot(\vec{x}-\vec{y})} e^{-iE_p\tau}\theta(\tau)$$
$$+ e^{-i\vec{p}\cdot(\vec{x}-\vec{y})} e^{iE_p\tau}\theta(-\tau) \}$$

The first term in the bracket leaves the volume integral $\int d^3p$ invariant when taking $p \to -p$.

To show this, consider only the terms with $\vec{p}\cdot(\vec{x}-\vec{y})$ in the exponential. The second term is just,

(i) $\int_{-\infty}^{+\infty} d^3 p e^{-i\vec{p}\cdot(\vec{x}-\vec{y})}$

Now take $p \to -p$, into the first term, we get

(ii) $\int_{+\infty}^{-\infty} d^3 (-p)e^{i(-\vec{p})\cdot(\vec{x}-\vec{y})} = \int_{+\infty}^{-\infty} -d^3 p e^{-i\vec{p}\cdot(\vec{x}-\vec{y})}$
$$= \int_{-\infty}^{+\infty} d^3 p e^{-i\vec{p}\cdot(\vec{x}-\vec{y})}$$

And so we have,

(12E.27) $\Delta(x - y) =$
$$\int \frac{d^3p}{(2\pi)^3(2E_p)} e^{-i\vec{p}\cdot(\vec{x}-\vec{y})} \{ e^{-iE_p\tau}\theta(\tau) + e^{iE_p\tau}\theta(-\tau) \}$$

We will evaluate the bracket term using the identity,

(12E.28) $\{ e^{-iE_p\tau}\theta(\tau) + e^{iE_p\tau}\theta(-\tau) \}$

$$= \lim_{\varepsilon \to 0} \frac{-2E_p}{2\pi i} \int_{-\infty}^{\infty} \frac{d\omega}{\omega^2 - E_p^2 + i\varepsilon} e^{i\omega\tau}$$

Proof:

We will do a contour integration. First we separate the denominator as partial fractions,

$$(12E.29) \quad \frac{1}{\omega^2 - E_p^2 + i\varepsilon} = \frac{1}{[\omega - (E_p - i\varepsilon)][\omega - E_p + i\varepsilon]}$$

$$= \frac{1}{2E_p} \left[\frac{1}{\omega - (E_p - i\varepsilon)} - \frac{1}{\omega - (-E_p + i\varepsilon)} \right]$$

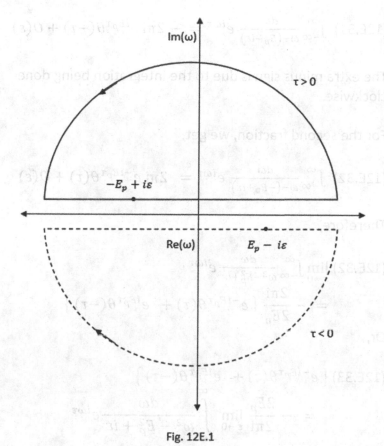

Fig. 12E.1

To do our integration we must deal with the Residue Theorem (appendix K, section KI), which states: Let f(z) be holomorphic, which depends on z but not on z^*. Let Γ be a closed anticlockwise contour in the complex plane. Then,

$$(12E.30) \quad \oint_\Gamma f(z) = 2\pi i \, \Sigma \text{ residues}$$

We have two poles around E_p.

The first fraction in 12E.29 picks up the pole only if $\tau < 0$,

(12E.31) $\int_{-\infty}^{\infty} \frac{d\omega}{\omega - (E_p - i\varepsilon)} e^{i\omega\tau} = -2\pi i \, e^{iE_p\tau}\theta(-\tau) + O(\varepsilon)$

The extra minus sign is due to the integration being done clockwise.

For the second fraction, we get,

(12E.32) $\int_{-\infty}^{\infty} \frac{d\omega}{\omega - (-E_p + i\varepsilon)} e^{i\omega\tau} = 2\pi i \, e^{-iE_p\tau}\theta(\tau) + O(\varepsilon)$

Therefore,

(12E.32) $\lim_{\varepsilon \to 0} \int_{-\infty}^{\infty} \frac{d\omega}{\omega^2 - E_p^2 + i\varepsilon} e^{i\omega\tau}$

$\qquad = -\frac{2\pi i}{2E_p} \left\{ e^{-iE_p\tau}\theta(\tau) + e^{iE_p\tau}\theta(-\tau) \right\}$

Or,

(12E.33) $\left\{ e^{-iE_p\tau}\theta(\tau) + e^{iE_p\tau}\theta(-\tau) \right\}$

$\qquad = -\frac{2E_p}{2\pi i} \lim_{\varepsilon \to 0} \int_{-\infty}^{\infty} \frac{d\omega}{\omega^2 - E_p^2 + i\varepsilon} e^{i\omega\tau}$

Putting it altogether by substituting 12E.33 into 12E.27,

(12E.34) $\Delta(x - y) =$
$\int \frac{d^3p}{(2\pi)^3(2E_p)} e^{-i\vec{p}\cdot(\vec{x}-\vec{y})} \left\{ e^{-iE_p\tau}\theta(\tau) + e^{iE_p\tau}\theta(-\tau) \right\}$

$$= \int \frac{d^3p}{(2\pi)^3 (2E_p)} e^{-i\vec{p}\cdot(\vec{x}-\vec{y})} \left\{-\frac{2E_p}{2\pi i} \lim_{\varepsilon\to 0} \int_{-\infty}^{\infty} \frac{d\omega}{\omega^2 - E_p^2 + i\varepsilon} e^{i\omega\tau}\right\}$$

Note that the factors $2E_p$ cancel each other, and $-\frac{1}{i} = i$

(12.E.35)

$$\Delta(x - y) = \lim_{\varepsilon\to 0} \int \frac{d^3p\, d\omega}{(2\pi)^4} \frac{i}{\omega^2 - E_p^2 + i\varepsilon} e^{-i\vec{p}\cdot(\vec{x}-\vec{y})} e^{i\omega\tau}$$

We can write using 12E.25,

(12E.36) $\omega^2 - E_p^2 + i\varepsilon = \omega^2 - [(|\vec{p}|)^2 + m^2] + i\varepsilon$

$$= \omega^2 - (|\vec{p}|)^2 - m^2 + i\varepsilon$$

$$= p^2 - m^2 + i\varepsilon$$

Only if we remember that $p = (\omega, \vec{p}) = (p^0, \vec{p})$ and $\omega = p^0 \neq E_p$

Also note that $d^3p\, d\omega = d^3p\, dp^0 = d^4p$

With this in mind, 12E.35 takes the form of,

(12E.37) $\Delta(x - y) = \int \frac{d^4p}{(2\pi)^4} \frac{i}{p^2 - m^2 + i\varepsilon} e^{-ip(x-y)}$

Note: when equation 12E.37 will be integrated, it will yield a complex number (a C-number).
The Fourier component of the Feynman propagator is then,

(12E.38) $\tilde{\Delta} = \frac{i}{p^2 - m^2 + i\varepsilon}$ (Appendix J, equation JA.4)

Note:

1) We assume that we had a free particle propagating between two space-time points, yet, $p^0 \neq E_p$, which suggests that we have a much more complicated situation when particles are allowed to create or annihilate.

Semantics: particles that obey equation 12E.25 are said to be on-shell; while particles in equation 12E.37 are said to be off-shell (Fig. 12E.2)

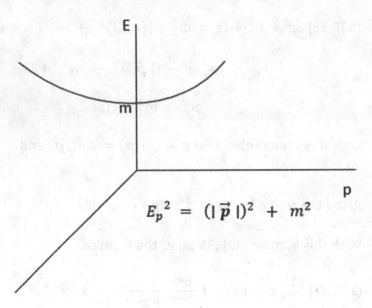

$$E_p^2 = (|\vec{p}|)^2 + m^2$$

Fig. 12E.2

2) Products of field operators such as found in equation 12E.1 do not occur in isolation but are parts of a process that involves the interaction picture, to which we will turn our attention in the following section.

12F. The Interaction Picture

First consider the case of no interaction – the free theory. We denoted the Hamiltonian as,

$$H_H(t) = H_S = H_O$$

We now consider the case that the interaction acts for a small amount of time.

The Hamiltonian can be broken up as,

$$H(t) = H_O + H'(t)$$

The idea is that the part with the free theory H_O is time-independent and can be easily solved. The part containing the interaction $H'(t)$ is time-dependent and we can use similar rules as in the Heisenberg picture.

We can now state the rules for the interaction picture as follows:

For any operator:

(12F.1) $O_I(t) = U^\dagger(t, 0) \, O \, U(t, 0)$ (equation 12B.4)

$$= e^{iH_0t} \, O \, e^{-iH_0t} \text{ (equation 12A.8)}$$

(12F.2) $i \frac{dO_I(t)}{dt} = [O_I(t), H_0]$ (equation12B.9)

For any quantum states:

(12F.3) $| \psi_I(t) >= e^{-iH_0t} | \psi(t) >$ (equation 12A.11)

(12F.4) $i \frac{d | \psi_I >}{dt} = H_I | \psi_I >$ (equation 12A.4)

Where

(12F.5) $H_I = e^{iH_0t} \, H' \, e^{-iH_0t}$ (equation 12F.1)

Applying the Schrödinger equation to the time-evolution operator, we have,

(12F.6) $i \frac{d}{dt_2} U_I(t_2, t_1) = H_I(t_2) U_I(t_2, t_1)$ (equation 12A.4)

Where $U_I(t, t) = 1$

We know the solution to equation 12F.6, however we must be careful that at different times, the Hamiltonian does not necessarily commute, that is, in general $[H_I(t_2), H_I(t_1)] \neq 0$. Now we bring back our time-ordered operator T and express our time-evolution as,

(12F.7) $U_I(t_2, t_1) = T \left[e^{-i \int_{t_1}^{t_2} dt H_I(t)} \right]$ (equ. 12A.10)

We can justify this as it satisfies equation 12F.6.

Proof:

We can treat $U_I(t_2, t_1)$ as a function of t_2. And take a derivative with respect to this time to get,

(12F.8) $\frac{d}{dt_2} T \left[e^{-i \int_{t_1}^{t_2} dt H_I(t)} \right] = T[-i H_I(t_2) e^{-i \int_{t_1}^{t_2} dt H_I(t)}]$

Recall that latest time operator to the left, since $H_I(t_2)$ is the latest time, we can pull it out of the time-ordered product, also multiply both sides by i, we get

(12F.9) $i \frac{d}{dt_2} T \left[e^{-i \int_{t_1}^{t_2} dt H_I(t)} \right] = H_I(t_2) T \left[e^{-i \int_{t_1}^{t_2} dt H_I(t)} \right]$

Comparing, we see that equation 12F.7 satisfies equation 12F.6.

QED

12G. The S-Matrix

In a scattering experiment, particles are fired at each other. Initially, we assume that they are free particles. So we can depict them as,

(12G.1) $|\phi> = |\phi_I(-\infty)>$

These are eigenstates of H_0. We can build them up from the vacuum state $|0>$. During the scattering process, we have H_I. We can say that the states have evolved in some ways, which is unknown to us. After this interaction, H_I is zero and the particles are free again. They are in some states denoted as,

(12G.2) $|\Psi> = |\Psi_I(+\infty)>$

We can write the expectation value as such,

(12G.3) $<\phi_I(0)|\Psi_I(0)> =$

$$<\phi_I(\infty)|U_I(\infty, 0)U_I(0, -\infty)|\Psi_I(-\infty)>$$

The RHS of the above equation is obtained by using equation 12A.1.

We can further contract by using equation 12A.3,

(12G.4) $<\phi_I(0)|\Psi_I(0)> =$

$$<\phi_I(\infty)|U_I(\infty, -\infty)|\Psi_I(-\infty)>$$

We define the S-matrix as,

(12G.5a) $S \equiv U_I(t, -t)$ as $t \to \infty$

(12G.5b) $<\phi_I(0)|\Psi_I(0)> = <\phi_I(\infty)|S|\Psi_I(-\infty)>$

Now that we have connected the S-matrix to the time evolution operator, we can rewrite equation 12F.7 as,

(12G.6) $S = U_I(\infty, -\infty) = T[e^{-i\int_{-\infty}^{\infty} d^4x H_I(x)}]$

Where we've replaced $\int dt\, H_I(t)$ by $\int d^4 H_I(x)$, using the Hamiltonian density.

We can Taylor expand (appendix E, equation EA.23) the exponential as,

(12G.7) $S = T[1 - i \int d^4 z\, H_I(z)$

$$+ \frac{(-i)^2}{2!} \int d^4 y\, d^4 w\, H_I(y)\, H_I(w) + \cdots]$$

The above is known as the Dyson's expansion.

Notice that each term of the expansion is made up of products of field operators. To deal with these objects, we will need an important theorem.

12H. Wick's Theorem

Consider the following product of field operators: <0|T[ABC...Z] |0 >. This is known as the Vacuum Expectation Value (VEV). An easier product of field operators to evaluate is the normal ordering operator, <0|N[ABC...Z] |0 >, in which all the annihilation operators are placed on the right, all the creation operators are placed on the left. Note that this product is zero.

From equation 11D.3 we write,

(12H.1) $\Phi(x) = \Phi^+(x) + \Phi^-(x)$,

(12H.2) where $\Phi^+(x) = \int d^3p(2\pi)^{-3} (2E_p)^{-1}a_p e^{-ip\cdot x}$, (1st term)

And $\Phi^+(x)$ annihilates, since $a_p|0> =0$

(12H.3) $\Phi^-(x) = \int d^3p(2\pi)^{-3}(2E_p)^{-1}a_p^+e^{ip\cdot x}$, (2nd term)

Where $\Phi^-(x)$ creates, since $a_p^+|0> =|1_p>$

Any product of two operators, AB, can be written as,

(12H.4) $AB = (A^+ + A^-)(B^+ + B^-)$

$$= A^+ B^+ + A^+ B^- + A^- B^+ + A^- B^-$$

Except for the third term, all the other terms have their operators in the normal ordering position. The trick is to add/subtract the term, $B^+ A^-$, which itself is in the normal ordering position,

(12H.5) $AB = A^+ B^+ + A^+ B^- + A^- B^+ + A^- B^- + (B^+ A^- - B^+ A^-)$

$$= A^+ B^+ + A^+ B^- + A^- B^- + B^+ A^- + (A^- B^+ - B^+ A^-)$$

Where we have rearranged so that all the normal ordering terms are gathered together, with the extra terms in a bracket, which is just a commutator. Therefore, we get the important result,

(12H.6) $AB = N(AB) + [A^-, B^+]$

We want to link this result to the time ordering operator. To do this, we must go back to the definition of the time ordering operator (equation 12E.2), that is, *later to the left*,

(12H.7) $T\{A(x) B(y)\}$

$$= A(x) B(y) \text{ for } x^0 > y^0, \text{ (} x^0 \text{ is later)}$$

$$= B(y) A(x) \text{ for } y^0 > x^0, \text{ (} y^0 \text{ is later)}$$

313

Applying this to equation 12H.6, we get,

(12H.8)

$$T\{A(x)\,B(y)\} = N\{A(x)B(y)\} + [A^-(x), B^+(y)], \text{ for } x^0 > y^0$$

$$= N\{A(x)B(y)\} + [B^-(x), A^+(y)], \text{ for } y^0 > x^0$$

Since the VEV of a normal ordered product is zero, we then have,

(12H.9)

$$< 0|T[A(x)B(y)]\,|0> = < 0|\,[A^-(x), B^+(y)]\,|0>, \text{ for } x^0 > y^0$$

$$= < 0|\,[B^-(x), A^+(y)]\,|0>, \text{ for } y^0 > x^0$$

If we choose $A = B = \Phi$, then the LHS is just the Feynman propagator (equation 12E.3), which is just a C-number

(equation 12E.37). We rewrite equation 12H.8 as,

(12H.10) $\overset{\sqcap}{AB} = T\{A(x)B(y)\} - N\{A(x)B(y)\}$

Where the lines above AB on the LHS denotes a contraction, which is a C-number. The normal ordering operator has no effect on C-number, only on operators. So, rearranging equation 12H.10 we get,

(12H.11) $T\{A(x)B(y)\} = N\{A(x)B(y)\} + \overset{\sqcap}{AB}$

$$= N\{A(x)B(y) + AB\}$$

We generalize this to any length of strings of field operators,

(12H.12) $T\{ABCDEF \ldots XYZ\} = N\{ABCDEF \ldots XYZ$

314

+all possible contractions of ABCDEF ... XYZ }

This is known as Wick's theorem.

Example of a string of four operators:

$$(12H.13)\ T\{ABCD\} = N\{ABCD\}$$

$$+N\{\overset{\sqcap}{A}\overset{}{B}CD\} + N\{A\overset{\sqcap\ \ \ }{B\ C}D\} + N\{AB\overset{\sqcap\ \ \ }{C\ D}\}$$

$$+N\{\overset{\sqcap\ \ \ \ }{A\ \ \ C}BD\} + N\{A\overset{\sqcap\ \ \ \ }{B\ \ \ D}C\} + N\{\overset{\sqcap\ \ \ \ \ \ }{A\ \ \ \ \ D}BC\}$$

$$+ N\{\overset{\sqcap}{A}\overset{\sqcap}{B}\overset{}{C}\overset{}{D}\} + N\{\overset{\ \sqcap\ \ }{A\ B\ C}D\} + N\{\overset{\sqcap\ \ \ \ }{A\ \ \ C}\overset{}{B}\overset{}{D}\}$$

Where in the 1st line, we have no contraction; in the 2nd and 3rd lines, we have one pair of contractions (6 possibilities); and in the 4th line, we have two pairs of contractions (3 possibilities). Now we can take advantage of this fact: every pair of field products gives a C-number which can then be pulled out of the product, since C-numbers commute with any operator. For instance for a single contraction we have,

(12H.14) N[ABCDEFGH...XYZ] = DH x N[ABCEFG...XYZ]

$$= \text{C-number } x\ N[ABCEFG...XYZ]$$

Where we've pulled the pair DH. But note as it was mentioned at the beginning of this section that, <0|N[ABC...Z] |0 > = 0. The only surviving terms are those for which all the operators have been paired off. In the above example, equation 12H.13 reduces to,

$$(12H.15)\ T\{ABCD\} = N\{\overset{\sqcap\ \sqcap}{A\ B\ C\ D}\} + N\{\overset{\ \sqcap\ \ }{A\ B\ C}\overset{}{D}\} + N\{\overset{\sqcap\ \ \ \ }{A\ \ \ C}\overset{}{B}\overset{}{D}\}$$

In the case of four bosonic operators (Let A = $\Phi(x_1)$, B = $\Phi(x_2)$, C= $\Phi(x_3)$, D =$\Phi(x_4)$), we have

(12H.16)

$$< 0 | T \, \Phi(x_1) \, \Phi(x_2) \, \Phi(x_3) \, \Phi(x_4) \, | 0 >$$

$$= \Delta(\, x_1 - x_2)\Delta(\, x_3 - x_4) + \Delta(\, x_1 - x_3)\Delta(\, x_2 - x_4)$$

$$+ \Delta(\, x_1 - x_4) \, \Delta(\, x_2 - x_3)$$

For instance in 12H.15 the first product of the first term is

→ AB → $\Phi(x_1) \, \Phi(x_2)$ → $\Delta(\, x_1 - x_2)$, which is the integral evaluated in 12F.37 as a C-number.

12I. Feynman's Diagrams

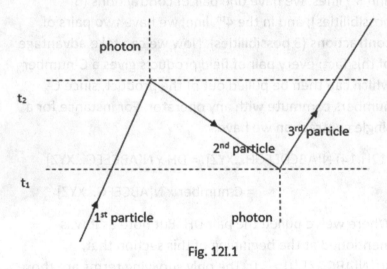

Fig. 12I.1

A typical Feynman's diagram contains the following (Fig. 12I.1):

(a) A particle is presented by a line with an arrow in the direction of time going up the page: Particles 1 and 3.
(b) An anti-particle is represented by a line with an arrow going in the opposite direction: Particle 2.

(c) Initially, we have a photon and a particle.

(d) At t_1, the photon gives way to the creation of a pair of particle/anti-particle: 2nd particle and 3rd particle

(e) At t_2, the anti-particle and the 1st particle interact and are annihilated to yield a photon.

The 5-step program for the ϕ^4 theory

We are one step closer to evaluate the Dyson equation (equation 12G.7). To do so, we will use the ϕ^4 theory as a practical illustration. In this theory the Lagrangian is modified by adding a term, the interaction. So consider the Klein-Gordon Lagrangian (equation 3D.3) with the ϕ^4 term,

12I.1) $\mathcal{L} = \frac{1}{2} (\partial_\mu \phi)^2 - \frac{m^2}{2} \phi^2 - \frac{\lambda}{4!} \phi^4$

The first two terms make up the free part of the theory,

(12I.2) $\mathcal{L}_0 = \frac{1}{2} (\partial_\mu \phi)^2 - \frac{m^2}{2} \phi^2$

And the interaction part is simply,

(12I.3) $\mathcal{L}_I = -\frac{\lambda}{4!} \phi^4$; $H_I = \frac{\lambda}{4!} \phi^4$

Step 1

The amplitude we need to calculate is in terms of the S-matrix (equations 12G.5b),

(12I.4) $A = <q \,|\, S \,|\, p>$

Where $|p>$ is the "in" state; and $<q|$ is the "out" state.

From equation 11F.6a-b, the above equation becomes,

(12I.5) $A = (2\pi)^3 (2E_p)^{1/2} (2E_q)^{1/2} < 0 \,|\, a_q S a_p^\dagger |\, 0 >$

Step 2

We rewrite the Dyson expansion, (12G.7), and substitute the interaction term (12I.3):

(12I.6) $S = T[\, 1 - i \int d^4 z H_I(z) +$

$\qquad\qquad \frac{(-i)^2}{2!} \int d^4 y d^4 w H_I(y) H_I(w) + \cdots]$

$= T\,[1 - i \dfrac{\lambda}{4!} \int d^4 z \; \phi(z)^4$

$\qquad\qquad + \frac{(-i)^2}{2!} (\frac{\lambda}{4!})^2 \int d^4 y \, d^4 w \; \phi(y)^4 \, \phi(w)^4 + \cdots]$

Step 3

Plug that into equation (12I.5),

(12I.7) $A = (2\pi)^3 (2E_p)^{1/2} (2E_q)^{1/2} T\,[< 0\,|\, a_q a_p^\dagger |\, 0 >$

$+ \left(\dfrac{-i\lambda}{4!}\right) \int d^4 z < 0 \,|\, a_q \phi(z)^4 a_p^\dagger |\, 0 >$

$+ \frac{(-i)^2}{2!} (\frac{\lambda}{4!})^2 \int d^4 y \, d^4 w \; < 0 \,|\, a_q \phi(y)^4 \phi(w)^4 a_p^\dagger |\, 0 >$

$+ \cdots]$

Notice that the above has the structure:

(12I.8) $A = A^{(0)} + A^{(1)} + A^{(2)} + \cdots$

Where each $A^{(n)}$ is proportional to λ^n.

Step 4

Using Wick's theorem, we begin with $A^{(1)}$. We notice there are two types of contractions.

(a) The first type contains contractions of the a-operator and the φ-operator separately.

For instance,

(12I.9)

$$< 0 \,|\, a_q \phi(z)\phi(z)\phi(z)\phi(z) a_p^\dagger \,|\, 0 >$$
$$= \,< 0 \,|\, a_q a_p^\dagger \,|\, 0 >< 0 \,|\,T\phi(z)\phi(z)\,|\, 0 >< 0 \,|\,T\phi(z)\phi(z)\,|\, 0 >$$

There are two more combinations of this type for a total of 3.

(b) The second type contains contraction between the a-operator and the φ-operator. There are twelve ways to contract. Here's one sample:

(12I.10)

$$< 0 \,|\, a_q \phi(z)\phi(z)\phi(z)\phi(z) a_p^\dagger \,|\, 0 >$$
$$= \,< 0 \,|\, a_q \phi(z)\,|\, 0 >< 0 \,|\,T\phi(z)\phi(z)\,|\, 0 >< 0 \,|\,\phi(z) a_p^\dagger \,|\, 0 >$$

Putting it altogether, we get from (12I.7),

(12I.11)
$$A^{(1)} = (2\pi)^3 (2E_p)^{\frac{1}{2}} (2E_q)^{\frac{1}{2}} \left(\frac{-i\lambda}{4!}\right) \int d^4 z$$

$$X \{3 < 0|a_q a_p^\dagger|0 >< 0|T\phi(z)\phi(z)|0 >< 0|T\phi(z)\phi(z)|0 >$$
$$+ 12 < 0\,|\,a_q\phi(z)\,|\,0 >< 0\,|T\phi(z)\phi(z)\,|\,0 >< 0\,|\,\phi(z)a_p^\dagger|\,0 >\}$$

Consider line 3 in the above equation. Recall from the example in 12H.16, we have the Feynman propagator in the second contraction,

$$(12I.12) \quad < 0|\ T\ \phi(z)\ \phi(z)\ |0 > =\ \Delta(\,z - z)$$

Also from line 3 in equation (12I.11) any contraction between a field and a creation operator corresponds to:

$$(12I.13) \quad < 0\,|\,\phi(z)a_p^\dagger|\,0 >$$
$$= \int \frac{d^3q}{(2\pi)^{\frac{3}{2}}} \frac{1}{(2E_q)^{\frac{1}{2}}} < 0\,|\,(a_q e^{-iq\cdot z} + a_q^\dagger e^{iq\cdot z})a_p^\dagger|\,0 >$$

Where we used equation 10A.14

Recall that $< 0|a_p^\dagger = 0$, so only 1st term of line 2 survives.

$$(12I.14) \quad < 0\,|\,\phi(z)a_p^\dagger|\,0 >$$
$$= \int \frac{d^3q}{(2\pi)^{\frac{3}{2}}} \frac{1}{(2E_q)^{\frac{1}{2}}} < 0\,|\,(a_q e^{-iq\cdot z})a_p^\dagger|\,0 >$$

Also, $a_p^\dagger|0 > = |\,p >$, $< 0|a_q =< q|$ and $< q\,|\,p > = \delta^{(3)}(\,q - p)$

$$(12I.15) \quad < 0\,|\,\phi(z)a_p^\dagger|\,0 >$$
$$= \int \frac{d^3q}{(2\pi)^{\frac{3}{2}}} \frac{1}{(2E_q)^{\frac{1}{2}}} (e^{-iq\cdot z})\delta^{(3)}(\,q - p)$$
$$= \frac{1}{(2\pi)^{\frac{3}{2}}} \frac{1}{(2E_p)^{\frac{1}{2}}} e^{-ip\cdot z} \text{ , after integrating over q.}$$

The full line 3 in equation 12I.11 with the outside factor now reads as,

(5I.15) $(2\pi)^3 (2E_p)^{\frac{1}{2}} (2E_q)^{\frac{1}{2}} \left(\frac{-i\lambda}{4!}\right) \int d^4 z$

$\qquad X(12) < 0 \,|\, a_q \phi(z)| \, 0 > < 0 \,|\, T\phi(z)\phi(z) \,|\, 0 > < 0 \,|\, \phi(z)a_p^\dagger| \, 0 >$

$\qquad = (2\pi)^3 (2E_p)^{\frac{1}{2}} (2E_q)^{\frac{1}{2}} \left(\frac{-i\lambda}{4!}\right) \int d^4 z$

$\qquad X(12) \dfrac{1}{(2\pi)^{\frac{3}{2}}} \dfrac{1}{(2E_p)^{\frac{1}{2}}} e^{iq\cdot z} \Delta(z-z) \dfrac{1}{(2\pi)^{\frac{3}{2}}} \dfrac{1}{(2E_p)^{\frac{1}{2}}} e^{-ip\cdot z}$

$\qquad = \left(\dfrac{-i\lambda}{2}\right) \int d^4 z \, e^{iq\cdot z} \Delta(z-z) e^{-ip\cdot z}$

Substitute the Feynman propagator (equation 12E.37) in the above,

(12I.16) $\left(\dfrac{-i\lambda}{2}\right) \int d^4 z \, e^{-iq\cdot z} \Delta(z-z) e^{-ip\cdot z}$

$= \left(\dfrac{-i\lambda}{2}\right) \int d^4 z \, e^{iq\cdot z} \dfrac{d^4 k}{(2\pi)^4} \dfrac{i}{k^2 - m^2 + i\varepsilon} e^{-ik(z-z)} e^{-ip\cdot z}$

$= \left(\dfrac{-i\lambda}{2}\right) \int d^4 z \, \dfrac{d^4 k}{(2\pi)^4} e^{iq\cdot z} \dfrac{i}{k^2 - m^2 + i\varepsilon} e^{-ip\cdot z}$

Step 5

We are now in the position of drawing the Feynman diagram.

Now going back to equations 12I.7 and 12I.8 we had

$A^{(1)} \rightarrow -\lambda \int d^4 z < 0 \,|\, a_q \phi(z)^4 a_p^\dagger| \, 0 >$ where this term contains one copy of $H_I(z)$.

So we draw four legs at position z.

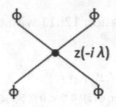

Fig. 12I.2

Recall in step 4, we had two types of contractions: 4a → a-operator and the ϕ-operator separately; 4b → between the a-operator and the ϕ-operator. We will consider doing 4b first as it contains most of the salient points in drawing the Feynman diagram. Repeating equation (12I.10) of step 4b below,

(12I.17)

$$< 0 \, | \, \overbrace{a_q \phi(z)}\overbrace{\phi(z)\phi(z)}\overbrace{\phi(z)a_p^\dagger} | \, 0 >$$
$$= \, < 0 \, | \, a_q \phi(z) | \, 0 > \, < 0 \, | \, T\phi(z)\phi(z) \, | \, 0 > \, < 0 \, | \, \phi(z)a_p^\dagger \, | \, 0 >$$

The contraction $\phi(z)a_p^\dagger$ is represented by an incoming line attached to one of the vertex.

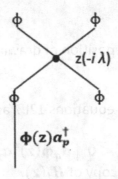

$$\phi(z)a_p^\dagger$$

Fig. (12I.3)

The contraction $\phi(z)\phi(z)$ ties together two vertex legs.

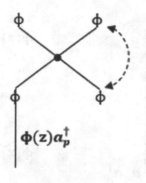

$\phi(z)a_p^\dagger$

Fig. (12I.4)

Similarly, the contraction $a_q\phi(z)$ is represented by an outgoing line attached to the remaining last leg of the vertex. We get the final result,

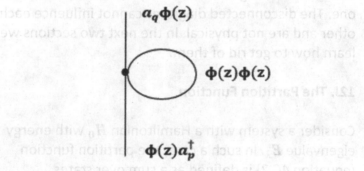

$a_q\phi(z)$

$\phi(z)\phi(z)$

$\phi(z)a_p^\dagger$

Fig. 12I.5

It can be verified that step 4a will give us two disconnected diagrams:

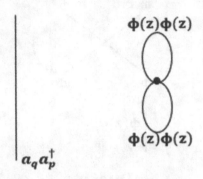

Fig. 12I.6

Notice that for first-order contributions we have a disconnected diagram while the previous was a connected one. The disconnected diagrams cannot influence each other and are not physical. In the next two sections we will learn how to get rid of them.

12J. The Partition Function

Consider a system with a Hamiltonian H_0 with energy eigenvalue E_λ. In such a case, the partition function (equation 4C.2) is defined as a sum over states,

$$(12J.1)\ Z = \sum_\lambda < \lambda \mid e^{-\beta H_0} \mid \lambda > = Tr\left[e^{-\beta H_0} \right]$$

Where $\beta = \frac{1}{k_B T}$, k_B is the Boltzmann constant and T is the temperature.

The probability of a system being in some particular state $\mid \lambda >$ with energy E_λ at a temperature T is given by,

(12J.2) $p_\lambda = \dfrac{e^{-\beta E_\lambda}}{Z}$

The thermal expectation value is found by summing up over the expectation values weighted by the probabilities of finding a state of energy E_λ.

$$(12J.3) \ <\varphi_i>_T = \Sigma_\lambda < \lambda \mid \varphi_i \mid \lambda > p_\lambda$$
$$= \frac{1}{Z}\Sigma_\lambda < \lambda \mid \varphi_i \mid \lambda > e^{-\beta E_\lambda}$$
$$= \frac{1}{Z}\Sigma_\lambda < \lambda \mid \varphi_i e^{-\beta H_0} \mid \lambda >$$
$$= \frac{Tr\,[\varphi_i\, e^{-\beta H_0}]}{Z}$$

A clever way to find the thermal average of a field is by adding a source term to the Hamiltonian, that is, $H_s = -\frac{1}{\beta}\Sigma_k J_k \varphi_k$. The partition function now reads as,

$$(12J.4) \ Z(J) = Tr\left[e^{-\beta H}\right]$$
$$= Tr\left[e^{-\beta(H_0+H_s)}\right]$$
$$= Tr\left[e^{-\beta H_0+\Sigma_k J_k \varphi_k}\right]$$

Now we can simply work out the thermal average $<\varphi_i>_T$ by differentiating $Z(J)$ with respect to J_i and evaluating at $J_i = 0$ and dividing by Z(J=0).

$$(12J.5) \ <\varphi_i>_T = \frac{1}{Z(J=0)}\left.\frac{\partial Z(J)}{\partial J_i}\right|_{J_i=0} = \frac{Tr\,[\varphi_i\, e^{-\beta H_0}]}{Z(J=0)}$$

We can generalize this to correlation of fields,

$$(12J.6) \ <\varphi_{i_1}\ldots\varphi_{i_n}>_T = \frac{1}{Z(J=0)}\left.\frac{\partial^n Z(J)}{\partial J_{i_1}\ldots\partial J_{i_n}}\right|_{J=0}$$

325

12K. Generating Functionals

Consider the Lagrangian with a source,

(12K.1) $\mathcal{L}[\varphi(x)] \rightarrow \mathcal{L}[\varphi(x)] + J(x)\varphi(x)$

We define the generating functional as:

(12K.2) $Z[J] \equiv\ <\ \Omega\ |\ U(\infty, -\infty)\ |\ \Omega >_J$

Where $U(\infty, -\infty)$ is the time-evolution operator of the full Hamiltonian and the J subscript means in the presence of a source J. The generating functional gives us the amplitude:

(12K.3)$Z[J] =<$ **no particles** $x^0 = \infty\ |$ **no particles** $y^0 = -\infty >_J$

The normalized generating functional is given as:

(12K.4) $\mathcal{Z}[J] = \dfrac{Z[J]}{Z[J=0]}$

This guarantees that $\mathcal{Z}[J = 0] = 1$, in other words, the amplitude which starts with no particle and ends with no particle in the absence of a source is unity.

Recall equation 12G.6, repeated below,

(12K.5) $U_I(\infty, -\infty) = T\left[e^{-i \int_{-\infty}^{\infty} d^4x H_I(x)} \right]$

With

(12K.6) $H_I = -J(x)\varphi_H(x)$

We need to specify that equation 12K.5 is defined in the Heisenberg picture. That is, we have

(12K.7) $Z[J] = \; < \Omega \,|\, T \left[e^{i \int_{-\infty}^{\infty} d^4 x J(x) \varphi_H(x)} \right] \,|\, \Omega >$

Taylor expanding (appendix E, equation EA.23),

(12K.8) $Z[J] =$

$$1 + \sum_{n=1}^{\infty} \frac{i^n}{n!} \int d^4 x_1 \ldots d^4 x_n \, J(x_1) \ldots J(x_n)$$
$$< \Omega \,|\, T \varphi_H(x_1) \ldots \varphi_H(x_n) \,|\, \Omega >$$

Recall that the Green function (equation 12E.13),

(12K.9) $G^{(n)}(x_1 \ldots x_n) = \; < \Omega \,|\, T \varphi_H(x_1) \ldots \varphi_H(x_n) \,|\, \Omega >$

Note that again this is in the Heisenberg picture. Differentiate equation 12K.8 as follows:

(12K.10) $G^{(n)}(x_1 \ldots x_n) = \left. \frac{1}{i^n} \frac{\delta^n Z[J]}{\delta J(x_1) \ldots J(x_n)} \right|_{J=0}$

Or

(12K.11) $G^{(n)}(x_1 \ldots x_n) = \left. \frac{1}{i^n} \frac{1}{Z[J=0]} \frac{\delta^n Z[J]}{\delta J(x_1) \ldots J(x_n)} \right|_{J=0}$

However, we want an equation for $Z[J]$ in terms of the interaction picture. So we start with an expression like equation 12K.7 but instead we write it in the interaction picture and acting on the vacuum state:

$$(12K.12)\ \mathcal{Z}[J] = \frac{Z[J]}{Z[0]} = \frac{<0|T\left[e^{-i\int_{-\infty}^{\infty}d^4x[H_I - J(x)\varphi_I(x)]}\right]|0>}{<0|T\left[e^{i\int_{-\infty}^{\infty}d^4xH_I}\right]|0>}$$

Similarly to what was done above (equation 5K.11), we differentiate the above equation n times, divide by i^n and set J=0:

$$(12K.13)\ G^{(n)}(x_1 \dots x_n)$$
$$= \frac{<0|T\varphi_I(x_1)\dots\varphi_I(x_n)\left[e^{-i\int_{-\infty}^{\infty}d^4x\ H_I}\right]|0>}{<0|T\left[e^{i\int_{-\infty}^{\infty}d^4xH_I}\right]|0>}$$

Recall equation 12G.6, repeated below,

$$(12K.14a)\ S = T\left[e^{-i\int_{-\infty}^{\infty}d^4xH_I(x)}\right]$$

We then have,

$$(12K.14b)\ G^{(n)}(x_1 \dots x_n) = \frac{<0|T\varphi_I(x_1)\dots\varphi_I(x_n)S|0>}{<0|S|0>}$$

Substitute equation 12K.9 in the LHS, we finally get what we were looking for,

$$(12K.15)\ <\Omega|T\varphi_H(x_1)\dots\varphi_H(x_n)|\Omega>$$
$$= \frac{<0|T\varphi_I(x_1)\dots\varphi_I(x_n)S|0>}{<0|S|0>}$$

This is known as the Gell-Mann—Low theorem.

Remarks:

(a) The state $|\Omega>$ is the ground state in the interaction picture with the full Hamiltonian H giving, $H|\Omega> = 0$.

(b) The field operators $\varphi_H(x)$ are in the Heisenberg picture.

(c) The RHS is written with $|0>$, the free ground state, defined as $H_0|0> = 0$, where $H = H_0 + H_I$.

In terms of the Feynman diagrams, the Gell-Mann—Low theorem gives us a simple result:

(12K.16) $< \Omega| T\varphi_H(x_1) ... \varphi_H(x_n)|\Omega> =$
$\sum$ *all connected diagrams with n external lines*

We get this results from the RHS, by dividing the numerator with the denominator:

(12K.17) Numerator $= < 0| T\varphi_I(x_1) ... \varphi_I(x_n)S|0>$
$=$
$\sum$ *all connected diagrams with n external lines*
          $X \exp(connected\ vacuum\ diagrams)$
(See step 5 in section 12I)

(12K.18) Denominator $= < 0|S|0>$
          $= \exp(connected\ vacuum\ diagrams)$

Now divide 12K.17 by 12K.18 and we get 12K.16.

12L. Feynman's Rules for a Scalar Field

EXAMPLE: We want to calculate the amplitude and probability of a particle A^+ annihilated by its anti-particle A^-

producing a spin-0 scalar boson B, which subsequently decays back into A⁺ and A⁻. The mass of the scalar boson is taken to be m_B.

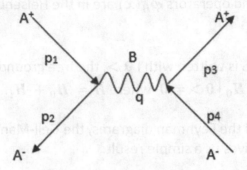

Fig. 12L.1

The Feynman rules for scalar fields are:
1. Write down one factor of $-i\lambda$ for each vertex where λ is the coupling constant of the interaction (Fig. 12I.1).

2. Add a propagator for each internal line (equation 12E.37).

3. Write a delta function for each vertex to conserve momentum.

In Fig. 12L.1, we have two vertices → $(-i\lambda)(-i\lambda) = -\lambda^2$

We have one internal line → $\int \frac{dq}{(2\pi)^4} \frac{i}{q^2 - m_B^2 + i\varepsilon}$

Two vertices means two delta functions →

$$(2\pi)^4\delta(p_1 + p_2 - q)(2\pi)^4\delta(q - p_3 - p_4)$$

The amplitude for this whole process:

(12L.1)

$$M = \int \frac{dq}{(2\pi)^4} \frac{-i\lambda^2}{q^2 - m_B{}^2 + i\varepsilon}$$
$$\times (2\pi)^4\delta(p_1 + p_2 - q)(2\pi)^4\delta(q - p_3 - p_4)$$

$$= \frac{-i\lambda^2}{(p_3 + p_4)^2 - m_B{}^2}$$

The probability is the square of the amplitude:

(12L.2) $P = \left(\dfrac{\lambda^2}{(p_3 + p_4)^2 - m_B{}^2}\right)^2$

Chapter 13

Path Integral Simplified

13A. Transition Amplitude

Consider a normalized wavefunction ψ,

(13A.1) $\psi = A_1\psi_1 + A_2\psi_2 + A_3\psi_3$

Where A_1 is the amplitude of ψ_1, A_2 is the amplitude of ψ_2, etc.

Given the rules of QM, the probability of measuring ψ_1 is

(13A.2) $P_1 = A_1 A_1^* = |A_1|^2$ (equation 7B.5)

The notion is that we started with ψ initially, then after measuring ψ_1, the wavefunction underwent a transition to a new state. Equation (13A.2) is the transition amplitude. Now, the transition amplitude in the canonical quantization approach of QFT is denoted by (equation 12G.5b),

(13A.3) for $t\pm \rightarrow \pm\infty$, $< f|U(t,t_0)|i > \equiv < f|S|i > \equiv S_{fi}$

Where $U(t,t_0) = \exp(-iH(t-t_0)/\hbar)$ is the time evolution operator (equation 12A.7).

However there is an elapsed time T between the measurements of the initial state ψ_i and the final state ψ_f. So we write,

(13A.4) $G(\psi_i,\psi_f;T) = <\psi_f|e^{-iHT/\hbar}|\psi_i>$, where G is the propagator (equation 12D.5).

13B. Wave Packets

If we take position as our eigenstates, then equation 13A.4 is,

(13B.1) $G(x_i,x_f;T) = <x_f|e^{-iHT/\hbar}|x_i> = <x_f|\psi>$

(13B.2) Where $|\psi> \equiv |e^{-iHT/\hbar}|x_i>$, is the evolved state

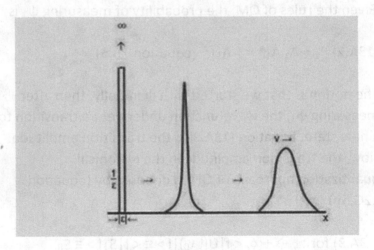

Fig. 13B.1

As the initial state evolves into ψ, like a wave packet it spreads and its peak diminishes (area is constant). The amplitude for measuring the particle at time T can be written as,

(13B.3) $G(x_i,x_f;T) = \int \delta(x - x_f)\psi(x,T)dx = \psi(x_f,T)$

The last part is from appendix J, equation JB.1. And

(13B.4) $|G(x_i,x_f;T)|^2 = \psi^*(x_f,T)\psi(x_f,T)$, is the probability density at x_f (equation 7B.5).

13C. Heuristic Argument

Writing the Schrödinger Equation (with $\hbar$ restored),

(13C.1) $i\hbar d|\psi\rangle/dt = H|\psi\rangle = E|\psi\rangle$

A general solution is,

(13C.2) $|\psi\rangle = C\, e^{-i(Et - \mathbf{p}\cdot\mathbf{x})/\hbar}$,

where C is a constant to be determined later on.

Now to simplify our notation, we will get rid of the Dirac notation, and simply write,

(13C.3) $\psi = Ce^{i\varphi}$

(13C.4) where $\varphi = -(Et - \mathbf{p}\cdot\mathbf{x})/\hbar$, is the phase angle and $e^{i\varphi}$ is the phasor.

The wave packet peak travels at the wave group velocity v, which corresponds to the classical particle velocity (Fig. 13B.1). The time rate of change of the phase angle at the peak is,

(13C.5) $d\varphi/dt = -(E - \mathbf{p} \cdot \mathbf{v})/\hbar$

(13C.6) But $E = T + V$, where T is the kinetic energy, and V is the potential energy (equation 2A.2).

(13C.7) Note: $T = \frac{1}{2}mv^2$ and $p = mv \rightarrow \mathbf{p} \cdot \mathbf{v} = 2T$

Substituting equations 13C.6 and 13C.7 into 13C.5, we get

(13C.8) $d\varphi/dt = (T - V)/\hbar = L/\hbar$, where L is the Lagrangian (equation 2A.1).

Integrating equation 13C.8,

(13C.9) $\varphi = \int Ldt/\hbar = S/\hbar$ where S is the action (equation 2A.4).

Therefore equation 13C.3 can now be written as,

(13C.10) $\psi = C(T)\exp(i \int Ldt/\hbar) = C(T) \exp(iS/\hbar)$,

where we take the more generally case that C might be time-dependent.

13D. Path Integral

The central idea of the Path Integral is that a particle/wave traveling between two events could be considered as traveling all possible paths between those two events.

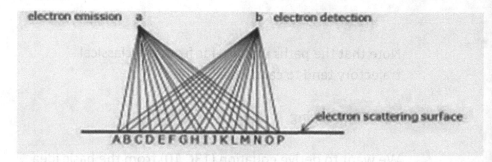

electron emission a   b electron detection

electron scattering surface

ABCDEFGHIJKLMNOP

Fig. 13D.1

In Fig. 13D.1, the Lagrangian is simply the kinetic energy of the electron, different for each path. These paths do not obey the usual classical laws – the least action principle, the equal angles of incidence and reflection law, and so on. But in the Path Integral, they must all be included. It turns out that if we calculate the phasor for each path we get the following (Fig.13D.2):

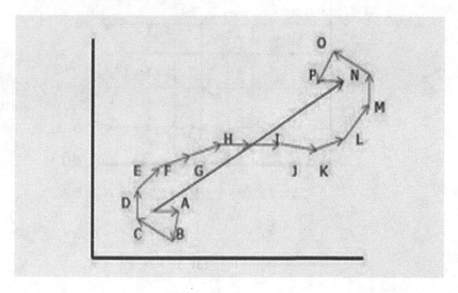

Fig. 13D.2

Note that the paths that are far from the classical trajectory tend to cancel each other.

13E. Time Slicing

We want to derive equation (13C.10) from the basic idea of the Path Integral. To do that we need to consider finite slices of time. We also discretize space and consider a small number of paths. In our example we will consider three different paths that we label as **a**, **b** and **c**, each of those over two time intervals, $t_0 \rightarrow t_1 \rightarrow t_2$.

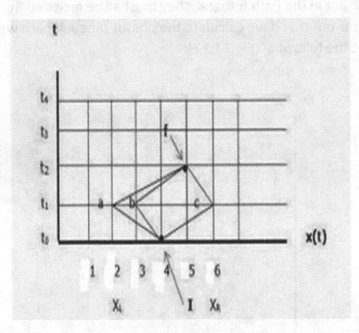

Fig. 13E.1

Recall that $L = T - V = \frac{1}{2}mv^2 - V(x)$

For path **a**:

(13E.1a) $L_{a1} = \frac{1}{2}m(x_{12}^2 - x_{04}^2)/\Delta t - V(\frac{1}{2}(x_{12} + x_{04}))$
(13E.1b) $L_{a2} = \frac{1}{2}m(x_{25}^2 - x_{12}^2)/\Delta t - V(\frac{1}{2}(x_{25} + x_{12}))$

For path **b**:

(13C.2a) $L_{b1} = \frac{1}{2}m(x_{13}^2 - x_{04}^2)/\Delta t - V(\frac{1}{2}(x_{13} + x_{04}))$
(13C.2b) $L_{b2} = \frac{1}{2}m(x_{25}^2 - x_{13}^2)/\Delta t - V(\frac{1}{2}(x_{25} + x_{13}))$

For path **c**:

(13C.3a) $L_{c1} = \frac{1}{2}m(x_{16}^2 - x_{04}^2)/\Delta t - V(\frac{1}{2}(x_{16} + x_{04}))$
(13C.3b)) $L_{c2} = \frac{1}{2}m(x_{25}^2 - x_{16}^2)/\Delta t - V(\frac{1}{2}(x_{25} + x_{16}))$

In terms of the phasors (equation 13C.10):

For path **a**:

(13C.4) $\exp(i/\hbar \int (L_{a1} + L_{a2})dt)$

$= \exp(i/\hbar \int L_{a1}dt) \exp(i/\hbar \int L_{a2}dt)$

$= \exp(i/\hbar \int \frac{1}{2}m(x_{12}^2 - x_{04}^2)/\Delta t - V(\frac{1}{2}(x_{12} + x_{04}))dt)$

$X \ \exp(i/\hbar \int \frac{1}{2}m(x_{25}^2 - x_{12}^2)/\Delta t - V(\frac{1}{2}(x_{25} + x_{12}))dt)$

$\approx \exp(iS(x_{04},x_{12}/\hbar) \exp(iS(x_{12},x_{25})/\hbar)$

We get similar expression for paths **b** and **c**.

(13C.5) The sum of these three paths

$$= \sum \exp(iS(x_{04},x_{1j})/\hbar) \exp(iS(x_{1j},x_{25})/\hbar), \, j=1,2,3$$

Since the transition amplitude is proportional to the above we can multiply by any constant, in this instance, C' and Δx_1, where $x_L < x_1 < x_R$.

(13C.6) $G(i,f;T) \approx C' \sum \exp(iS(x_{04},x_{1j})/\hbar) \exp(iS(x_{1j},x_{25})/\hbar)\Delta x_1$

$$\approx C' \int_{x_L}^{x_R} dx_1 \exp(iS(x_{04},x_1)/\hbar) \exp(iS(x_1,x_{25})/\hbar)$$

$$\approx C' \int_{x_L}^{x_R} \exp[i \int_{x_{04}}^{x_{25}} (L/\hbar)dt] \, dx_1$$

In the general case, we have intervals dx_1, dx_2, dx_3 ... dx_N, where $N \rightarrow \infty$

(13C.7) $G(i,f;T=t_f-t_i) = C \iiint ... \int e^{i\int \left(\frac{L}{\hbar}\right)dt} \, dx_1 dx_2 dx_3 ... \, dx_n$

$$= C(T) \int e^{i\int \left(\frac{L}{\hbar}\right)dt} \, Dx,$$

Where the symbol D implies all paths between i and f.

QED

13F. Derivation of the Constant

Consider equation 13B.1 rewritten below,

(13F.1) $G(x_i,x_f;T) = < x_f | e^{-iHT/\hbar} | x_i >$

The bra and ket are Dirac delta functions (equation 7B.4),

(13F.2) $G(x_i,x_f;T) = \int \delta(x - x_f) e^{-iHT/\hbar} \delta(x - x_i) dx$

Mathematically the Dirac Delta function is (equation JB.5):

(13F.3) $\delta(x - x_i) = (2\pi)^{-1} \int e^{ik(x - x_i)} \, dk$, (for the ket)

$$= (2\pi\hbar)^{-1} \int e^{ip(x - x_i)} \, dp$$

The second line is obtained with $p = \hbar k$

Similarly for the bra,

(13F.4) $\delta(x - x_f) = (2\pi\hbar)^{-1} \int e^{ip'(x_f - x)} \, dp'$

When we substitute equation 13F.3 and 13F.4 into 13F.2, we will get three exponential functions, and a triple integral. To simplify matters, we will examine the argument of each of the exponential function separately, from left to right, and then combine, taking care of not changing the order since we are dealing with operators.

(13F.5) For the bra, the exponent $E_1 = (ip/\hbar)(x - x_i)$

(13F.6) For the factor $e^{-iHT/\hbar}$, $E_2 = -iHT/\hbar = -iET/\hbar$

Where operating on the initial state, $H = E$, which is the eigenvalue. Note: we have a number (E) instead of an operator (H).

(13F.7) For the ket, the exponent $E_3 = (ip'/\hbar)(x_f - x)$

Now we can pass E_2 as it is a number. So, we can write it as $\exp(E_2) \exp(E_1) \exp(E_3)$.

Let us examine the last two exponents:

(13F.8) $\exp(E_1)\exp(E_3)$

$$= e^{(\frac{ip}{\hbar})(x-x_i)}e^{(\frac{ip'}{\hbar})(x_f-x)} = e^{i(p-p')x/\hbar}e^{ip'x_f/\hbar}e^{-ipx_i/\hbar}$$

The first exponent in equation 13F.8, combine with $(2\pi\hbar)^{-1}\int dx$ to give,

(13F.9) $(2\pi\hbar)^{-1}\int e^{i(p-p')x/\hbar}\,dx = \delta(p-p')$

With this result, equation 13F.2 becomes,

(13F.10) $G(x_i,x_f;T)$

$$= (2\pi\hbar)^{-1}\int\int e^{-\frac{iET}{\hbar}}\delta(p-p')e^{ip'x_f/\hbar}e^{-ipx_i/\hbar}dpdp'$$

$$= (2\pi\hbar)^{-1}\int e^{-\frac{iET}{\hbar}}e^{(\frac{ip}{\hbar})(x_f-x_i)}dp$$

$$= (2\pi\hbar)^{-1}\int e^{-\frac{ip^2T}{2m\hbar}}e^{(\frac{ip}{\hbar})(x_f-x_i)}dp$$

Where the energy is kinetic, that is,

(13F.11) $E = p^2/2m$.

Using appendix I, (equation IC.7),

(13F.12) $\int_{-\infty}^{\infty} dy\, e^{a^2y+by} = (\pi/a)^{\frac{1}{2}}e^{b^2/4a}$

Making the following correspondence,

$y \to p,\ a \to iT/2m\hbar,\ b \to (i/\hbar)(x_f-x_i)$

342

(13F.13) $G(x_i, x_f; T) = (m/(i2\pi\hbar T))^{\frac{1}{2}} e^{im(x_f - x_i)^2/(2T\hbar)}$

The probability density is then

(13F.14) $|G(x_i, x_f; T)|^2 = m/(2\pi\hbar T)$

We can deduce from equation (13F.14) that:

(i) For very large T, the probability amplitude decreases. That means that the farther away the path is from the classical path (the longer time it will take to go from x_i to x_f), the less it contributes to the probability amplitude (Fig. 13D.2).

(ii) As the mass m increases, so is the height, thus the width must decrease (area under envelope is constant - Fig. 13B.1) That is, the wave packet approaches the classical behavior of a particle.

(iii) If $\hbar$ were to go to zero, the peak would be infinite, giving an exact location, as it should for a classical particle (Fig. 13B.1).

Going back to the central idea of the Path Integral - which is that a particle/wave traveling between two events could be considered as traveling all possible paths between those two events - we now see that the greater deviation from the classical path, the smaller contribution we get to the probability amplitude. In addition, we also find that for large mass, or $\hbar = 0$, we fall into the classical regime.

Chapter 14

Spontaneous Symmetry Breaking

The following is a general review.

14A. Gauge Theory in QFT

(14A.1) in QM: x → operator

     in QFT: x → parameter, and $\phi(x)$ → operator.

Now $\phi(x)$, a function of x, is called the "field".

(14A.2) L = T − V. The Lagrangian plays an important role. Corresponding to L there is a Hamiltonian, H = T + V. The Hamiltonian is known to measure the energy of a system.

(14A.3) In classical mechanics, we have v = dx/dt, and L = ½ mv² − V(x).

The corresponding Hamiltonian is, H = ½ mv² + V(x). Quantizing this, ($\hbar$ =1),we get the Schrödinger equation (equation 7C.14):

$i\partial\Psi(x)/\partial t =(-\frac{1}{2}m\nabla^2 + V(x))\Psi(x).$

(14.4) In Relativity, the energy equation is:

$E^2 = p^2c^2 + m^2c^4.$

Quantizing this, (c =1) yields the K-G equation (10A.10a):

(14A.5) $(\partial_\mu\phi)(\partial^\mu\phi) + m^2\phi^2 = 0.$

From this, the Lagrangian can be deduced as:

$L = \frac{1}{2}(\partial_\mu\phi)^2 - \frac{1}{2}m^2\phi^2$.

(14A.6) In QFT, the general Lagrangian is:

$L = \frac{1}{2}(\partial_\mu\phi)^2 - V(\phi)$.

Comparing (14A.6) and (14A.5), if $V(\phi)$ contains any terms with ϕ^2, its coefficient is taken to be the mass of the field quanta (particles).

(14A.7) $V(\phi) \rightarrow kV(\phi) \rightarrow k\phi^2 \rightarrow k = \frac{1}{2}m^2$

Or $m = \sqrt{2k}$

14B. Abelian Gauge Theory

From electromagnetism, it was known that Maxwell's equations were gauge invariant. In QM, gauge invariance of the Lagrangian involves three important steps:

(14B.1) the wave function is transformed as $\phi \rightarrow e^{iq\chi}\phi$

(14B.2) the operator $\partial_\mu \rightarrow \partial_\mu + iqA^\mu$

(14B.3) the electromagnetic field $A^\mu \rightarrow A^\mu - \partial^\mu\chi$

(14B.4) In QED, that is Quantum Electrodynamics, the quantum version of electromagnetism, we have

$V(\Phi) \rightarrow - \frac{1}{4} F_{\mu\nu} F^{\mu\nu}$, where $F^{\mu\nu} = \partial^\mu A^\nu - \partial^\nu A^\mu$, (see equations 3G.3, 3G.5 and 3G.15).

If you apply equations 14B.1-4 to equation 14A.5, you get the invariance of the Lagrangian under gauge transformation, in which the photon mediates the electromagnetic force. Note that the photon has no mass.

In the weak force, it was known that the bosons involved in mediating that force have mass, and one had to figure out how to include a mass term, keeping the Lagrangian gauge invariant.

14C. Higgs Mechanism

Basically, we will only look at a U(1) symmetry for simplicity and see how mass is introduced in the Lagrangian of equation (14A.6).

Rewriting this equation as:

(14C.1) $L = \partial_\mu \Phi^\dagger \partial^\mu \Phi - \frac{1}{4} F_{\mu\nu} F^{\mu\nu} - V(\Phi^\dagger \Phi)$.

Where the second term refers to the electromagnetic interaction (equation 3G.15). And where the third term is,

(14C.2) $V(\Phi^\dagger \Phi) = (m^2)/(2\phi^2) \{\Phi^\dagger \Phi - \phi^2\}^2$

Three important things to note by this introduction:

(14C.3) The field Φ is now a complex number, denoted by

(Φ_1, Φ_2) or $\Phi = \Phi_1 + i\Phi_2$, and $\Phi^\dagger = \Phi_1 - i\Phi_2$ (see section 11G).

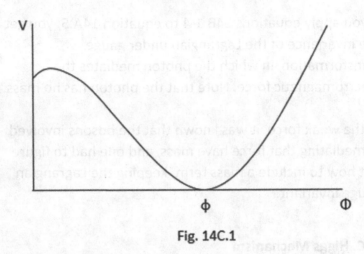

Fig. 14C.1

(14C.4) The minimum field energy is obtained when $\Phi^\dagger\Phi = \phi^2$, the vacuum state. But this implies there are many states in the space (Φ_1, Φ_2) as we can transform the state around a circle $|\Phi^2| = $ constant, which is a U(1) symmetry (see appendix H, section HC).

(14C.5) Since the number of possible vacuum states is infinite we need to break this symmetry by requiring that Φ is real, we take the vacuum state to be $(\phi,0)$, and expand: $\Phi = \phi + \frac{1}{\sqrt{2}}h$.

This is what is to be understood as spontaneous symmetry breaking.

Substituting equations 14B.1-3, equations 14C.2, 14C.5 into 14C.1, we get

(14C.6) $L = \{(\partial_\mu - iqA_\mu)(\phi + \frac{1}{\sqrt{2}}h)\}\{(\partial^\mu + iqA^\mu)(\phi + \frac{1}{\sqrt{2}}h\}$

$$- \tfrac{1}{4} F_{\mu\nu} F^{\mu\nu} - (m^2)/(2\phi^2) \{\sqrt{2}\phi h + \tfrac{1}{2}h^2\}^2$$

After calculating the Lagrangian, we separate it into two parts:

(14C.7) $L = L_{free} + L_{int}$

Where

(14C.8a) $L_{free} \equiv \tfrac{1}{2}\partial_\mu h \partial^\mu h - m^2 h^2 - \tfrac{1}{4} F_{\mu\nu} F^{\mu\nu} + q^2\phi^2 A_\mu A^\mu$

(14C.8b) $L_{int} \equiv q^2 A_\mu A^\mu (\sqrt{2}\phi h + \tfrac{1}{2}h^2)$

$$- (m^2 h^2/2\phi^2)(\sqrt{2}\phi h + \tfrac{1}{4}h^2)$$

So using equation 14A.7, we can see that by breaking the symmetry, we end up with two *__massive__* particles. In equation 14C.8a, the second term refers to a scalar particle with mass equal to $\sqrt{2}m$, associated with h (the Higgs field) and the fourth term, a vector boson with mass $\sqrt{2}q\phi$, associated with A^μ (the electromagnetic field). In the electroweak theory, there was a need for a vector boson with mass. The Higgs mechanism does that by introducing the Higgs field.

Appendix A

Greek Alphabet

Letter	Name	Letter	Name
A, α	alpha	N, ν	nu
B, β	beta	Ξ, ξ	ksi
Γ, γ	gamma	O, o	omicron
Δ, δ	delta	Π, π	pi
E, ε	epsilon	P, ρ	rho
Z, ζ	zeta	Σ, σ	sigma
H, η	eta	T, τ	tau
Θ, θ	theta	Y, υ	upsilon
I, ι	iota	Φ, φ	phi
K, κ	kappa	X, χ	chi
Λ, λ	lambda	Ψ, ψ	psi
M, μ	mu	Ω, ω	omega

Appendix B

Dimensional Units

Quantity	Symbol	$c = 1$	$\hbar = c = 1$
Mass	M	M	M
Length	L	L	M^{-1}
Time	T	L	M^{-1}
Velocity	L/T	dimensionless	dimensionless
Momentum	ML/T	M	M
Force	ML/T^2	M/L	M^2
Energy	ML^2/T^2	M	M

In the SI units, [Mass] = kilogram (kg),

[Length] = meter (m)
[Time] = second (s)
[Force] = Newton (nt = $kgms^{-2}$)
[Energy] = Joule (J = kgm^2s^{-2})
[Charge] = Coulomb (C)
[Voltage] = Volt (V = J/C)

Universal Constants: $c = 3 \times 10^8$ ms^{-1}
$\hbar = 1 \times 10^{-34}$ Js
$G = 6.67 \times 10^{-11}$ $m^3kg^{-1}s^{-2}$
$e = 1.6 \times 10^{-19}$ C

In natural units ($\hbar = c = 1$), mass, momentum and energy have the same dimension. It is customary to measure these in MeV (c=1). An electronvolt (eV) is an energy unit defined as 1.6×10^{-19} J. One MeV = 10^6 eV, one GeV = 10^3 MeV.

The mass equivalent in electronvolts ($E = mc^2$) is then:

$$m = \frac{E}{c^2} = \frac{1.6 \times 10^{-19} \frac{J}{eV}}{(3 \times 10^8 ms^{-1})^2}$$

$$= 1.78 \times 10^{-36} \frac{kgm^2s^{-2}}{m^2s^{-2}}/eV$$
$$= 1.78 \times 10^{-36} kg/eV$$

Example: Mass of an electron = $9.1 \times 10^{-31} kg$
$$= \frac{9.1 \times 10^{-31} kg}{1.78 \times 10^{-36} kg/eV}$$
$$= 5.11 \times 10^5 eV$$
$$= .511 \text{ MeV}$$

Example: Mass of a proton = $1.67 \times 10^{-27} kg$
$$= \frac{1.67 \times 10^{-27} kg}{1.78 \times 10^{-36} kg/eV}$$
$$= 0.938 \times 10^9 eV$$

$$= 938 \text{ MeV}$$

Often the mass of a proton is quoted as ≈ 1 GeV and is approximately 2000 times the mass of an electron.

Appendix C

Functions

CA. Important Summation Expansions

(CA.1) $\dfrac{1}{1-k} = 1 + k + k^2 + k^3 + \ldots$

(CA.2) $e^x = 1 + x + \dfrac{x^2}{2!} + \dfrac{x^3}{3!} + \ldots$

where $e \equiv 2.71828\ldots$ is a special number that plays an important role in calculus (see appendix C, equation CD.8).

CB. Polar Coordinates

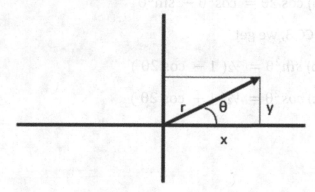

Fig. CB.1

(CB.1) $x = r \cos \theta$

(CB.2) $y = r \sin \theta$

(CB.3) $r^2 = x^2 + y^2$

This is also known as Pythagoras Theorem.

CC. Trigonometric relations

(CC.1) $\tan \theta = \dfrac{\sin \theta}{\cos \theta} = \dfrac{y}{x}$

(CC.2) $\csc \theta = \dfrac{1}{\sin \theta}$, $\sec = \dfrac{1}{\cos \theta}$, $\cot \theta = \dfrac{1}{\tan \theta}$

(CC.3) $\cos^2 \theta + \sin^2 \theta = 1$

(CC.4) $\sec^2 \theta - \tan^2 \theta = 1$

(CC.5) $\csc^2 \theta - \cot^2 \theta = 1$

(CC.6) $\sin (\theta \pm \varphi) = \sin \theta \cos \varphi \pm \cos \theta \sin \varphi$

(CC.7) $\cos (\theta \pm \varphi) = \cos \theta \cos \varphi \mp \sin \theta \sin \varphi$

For $\theta = \varphi$ in the above sum, we get

(CC.7a) $\cos 2\theta = \cos^2 \theta - \sin^2 \theta$

Using CC.3, we get

(CC.7b) $\sin^2 \theta = \frac{1}{2}(1 - \cos 2\theta)$

(CC.7c) $\cos^2 \theta = \frac{1}{2}(1 + \cos 2\theta)$

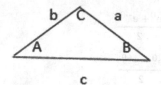

Fig. CC.1

(CC.8) Sine Law: $\dfrac{a}{\sin(A)} = \dfrac{b}{\sin(B)} = \dfrac{c}{\sin(C)}$

(CC.9) Cosine Law: $c^2 = a^2 + b^2 - 2ab\cos(C)$

CD. Exponentials and Logarithms

(CD.1) $a^m\, a^n = a^{m+n}$

(CD.2) $\dfrac{a^m}{a^n} = a^{m-n}$

(CD.3) $y = a^x \rightarrow x = \log_a y$,

where x is the exponent of a that yields y.

(CD.4) $\log_a(xy) = \log_a x + \log_a y$

(CD.5) $\log_a(x/y) = \log_a x - \log_a y$

(CD.6) $\log_a(y)^n = n\log_a y$

(CD.7) $\log_a b = \dfrac{\log_a y}{\log_b y}$

(CD.8) The natural logarithm is denoted by $\ln x \equiv \log_e x$,

(CD.9) $e^{i\theta} = \sin 0 + i\cos\theta$

CE. Hyperbolic Functions

(CE.1) $\sinh \theta = \dfrac{e^{\theta} + e^{-\theta}}{2}$

(CE.2) $\cosh \theta = \dfrac{e^{\theta} - e^{-\theta}}{2}$

(CE.3) $\tanh \theta = \dfrac{\sinh \theta}{\cosh \theta}$

(CE.4) $\operatorname{csch} \theta = \dfrac{1}{\sinh \theta}$, $\operatorname{sech} = \dfrac{1}{\cosh \theta}$, $\coth \theta = \dfrac{1}{\tanh \theta}$

(CE.5) $\cosh^2 \theta - \sinh^2 \theta = 1$

$\rightarrow 1 - \dfrac{\sinh^2 \theta}{\cosh^2 \theta} = \dfrac{1}{\cosh^2 \theta}$

$\rightarrow 1 - \tanh^2 \theta = \dfrac{1}{\cosh^2 \theta}$

$\rightarrow \cosh^2 \theta = \dfrac{1}{1 - \tanh^2 \theta}$

(CE.6)$\rightarrow \cosh \theta = \sqrt{\dfrac{1}{1 - \tanh^2 \theta}}$

(CE.7) $\operatorname{sech}^2 \theta + \tanh^2 \theta = 1$

(CE.8) $\coth^2 \theta - \operatorname{csch}^2 \theta = 1$

(CE.9) $e^{\theta} = \cosh \theta + \sinh \theta$

(CE.10) $e^{-\theta} = \cosh \theta - \sinh \theta$

CF. Binomial Coefficient

The binomial coefficient is the coefficient of the x^k in the polynomial $(1 + x)^n$, and is given by $(n \geq k \geq 0)$:

(CF.1) $\dbinom{n}{k} = \dfrac{n!}{k!(n-k)!}$

Example: Consider $(1 + x)^4$

$$= \binom{4}{0} x^0 + \binom{4}{1} x^1 + \binom{4}{2} x^2 + \binom{4}{3} x^3 \binom{4}{4} x^4$$

The binomial coefficient of x^2:

$$\binom{4}{2} = \frac{4!}{2!\,(4-2)!} = \frac{4!}{2!\,(2)!} = \frac{4 \cdot 3 \cdot 2 \cdot 1}{2 \cdot 1 \cdot (2 \cdot 1)} = 6$$

For the full expression we get:

$$(1 + x)^4 = 1 + 4x + 6x^2 + 4x^3 + x^4$$

Appendix D

Linear Algebra

Vectors are denoted by a (boldfaced) **V**, or an upper index (V^i) or lower index (V_i), where Latin index $i = 1,2,3$. In the 4-vector formalism, a Greek index $\mu = 0,1,2,3$ is used. Sometimes we expressed explicitly the vectors in terms of its components: $\mathbf{V} = V_i = (V_x, V_y, V_z)$

DA. Vector Analysis

Two important rules for vectors. We're working in 2D for simplicity:

$$\text{(DA.1)}\ A + B = \left(A_x i + A_y j\right) + \left(B_x i + B_y j\right)$$

$$= (A_x + B_x)i + (A_y + B_y)j$$

This shows that one can add components independently.

(DA.2) $c(A + B) = c(A_x i + A_y j) + c(B_x i + B_y j)$

$$= c(A_x + B_x)i + c(A_y + B_y)j$$

Multiplying vectors by a constant c lengthens the vector (for c > 1), or shortens it (c < 1).

Two major operations between vectors (in 3D):

(DA.3a) Dot product: $(\mathbf{V \cdot V})$ = $V_i V_i = V^2 = V_x{}^2 + V_y{}^2 + V_z{}^2$

Note: the Einstein summation means that for any repeated pair of indices we have an implied sum, that is, $V_i V_i = V_1 V_1 + V_2 V_2 + V_3 V_3$, where I = 1,2,3.

In DA.3a we use $i \cdot i = j \cdot j = k \cdot k = 1$;

and $i \cdot j = j \cdot k = k \cdot i = 0$

(DA.3b) $|a \cdot b|$ = ab cos θ

Proof:

Consider Fig. DA.1. In triangle OAB, we have three vectors: **a**, **b** and **a-b**. The angle θ is between segments OB and OA. Applying the law of cosines (equation CC.9), we have

(i) → $|a - b|^2 = |a|^2 + |b|^2 - 2|a||b| \cos θ$

(ii) Also → $|a - b|^2 = (a - b) \cdot (a - b)$

$= a \cdot a - a \cdot b - b \cdot a + b \cdot b$

$= |a|^2 + |b|^2 - 2|a \cdot b|$

Comparing (i) and (ii), we have,

$|\mathbf{a} \cdot \mathbf{b}| = ab \cos$

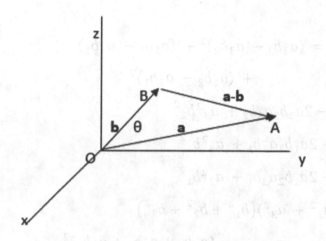

Fig. DA.1

(DA.4a) The second operation is the cross product:

$$\mathbf{a} \times \mathbf{b} = \begin{vmatrix} i & j & k \\ a_1 & a_2 & a_3 \\ b_1 & b_2 & b_3 \end{vmatrix}$$

Where we take the determinant (see below equation DB.8b).

$$= (a_2 b_3 - a_3 b_2)i - (a_1 b_3 - a_3 b_1)j + (a_1 b_2 - a_2 b_1)k$$

(DA.4b) Note: $\mathbf{a} \times \mathbf{a} = 0$

Also, if we are interested in any of the components, we can readily find it by using (z-component):

(DA.4c) $[\mathbf{A} \times \mathbf{B}]_z = A_x B_y - A_y B_z$

Take cyclic term: $x \rightarrow y, y \rightarrow z, z \rightarrow x$ for the other components.

(DA.4d) $|\mathbf{a} \times \mathbf{b}| = ab \sin \theta$

Proof:

$\rightarrow |\mathbf{a} \times \mathbf{b}|^2 = (a_2 b_3 - a_3 b_2)^2 + (a_1 b_3 - a_3 b_1)^2$
$$+ (a_1 b_2 - a_2 b_1)^2$$

$= a_2{}^2 b_3{}^2 - 2a_2 b_3 a_3 b_2 + a_3{}^2 b_2{}^2$

$+ a_1{}^2 b_3{}^2 - 2a_1 b_3 a_3 b_1 + a_3{}^2 b_1{}^2$

$+ a_1{}^2 b_2{}^2 - 2a_1 b_2 a_2 b_1 + a_2{}^2 b_1{}^2$

$= (a_1{}^2 + a_2{}^2 + a_3{}^2)(b_1{}^2 + b_2{}^2 + b_3{}^2)$
$$-(a_1 b_1 + a_2 b_2 + a_3 b_3)^2$$

$= |\mathbf{a}|^2 |\mathbf{b}|^2 - |\mathbf{a} \cdot \mathbf{b}|^2$

$= a^2 b^2 - a^2 b^2 \cos^2\theta = a^2 b^2 (1 - \cos^2\theta) = a^2 b^2 \sin^2\theta$

Where we used CC.3.

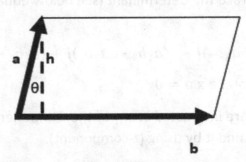

Fig. DA.2

Note that the cross product gives the area of a parallelogram, Fig. DA.2.

→ Area = Height X Width = bh = ab sin θ

Special operators:

(DA.5) Divergence: $\nabla V = \dfrac{\partial V_x}{\partial x} + \dfrac{\partial V_y}{\partial y} + \dfrac{\partial V_z}{\partial z}$

$$= \partial_x V_x + \partial_y V_y + \partial_z V_z$$

(DA.6) Curl: $[\nabla \times B]_z = \partial_x B_y - \partial_y B_z$

Again, take cyclic term: $x \to y, \ y \to z, \ z \to x$

DB. Matrices

An m x n matrix $A = (A_{ij}; i = 1, \ldots, m; j = 1, \ldots, n)$ is represented as:

(DB.1) $A = \begin{pmatrix} A_{11} & A_{12} & \cdots & A_{1n} \\ A_{21} & A_{22} & \cdots & A_{2n} \\ \cdots & \cdots & \cdots & \cdots \\ A_{m1} & A_{m2} & \cdots & A_{mn} \end{pmatrix}$

The complex conjugate of **A**, written as **A***, is defined as

(DB.2) $A^* = (A_{ij}^*)$

The transpose of A, written as A^T, is defined as

(DB.3) $A_{ij}^T = A_{ji}$

The Hermitian conjugate of **A,** written as $A^\dagger$, is defined as

(DB.4) $A_{ij}^{\dagger} = A_{ij}^{T*} = A_{ji}^{*}$

Multiplication of matrices: multiply a row from one matrix to the column of the second matrix, element by element and add the result:

$$(DB.5)\quad \begin{pmatrix} A & B \\ C & D \end{pmatrix}\begin{pmatrix} E & F \\ G & H \end{pmatrix} = \begin{pmatrix} AE + BG & AF + BH \\ CE + DG & CF + DH \end{pmatrix}$$

The first element of the product is $AE + BG$. This is obtained by multiplying the elements of the first row, that is (A B) of the first matrix, with the elements of the first column $\begin{pmatrix} E \\ G \end{pmatrix}$ in the second matrix. So A with E, B with G, then add them.

(DB.6) The trace of **A,** written as Tr **A,** is defined as

Tr **A** $= A_{ii}$

This is the sum of the diagonal elements.

Example: A = $\begin{pmatrix} a & 0 & 0 & 0 \\ 0 & -b & 0 & 0 \\ 0 & 0 & c & 0 \\ 0 & 0 & 0 & 4 \end{pmatrix}$

Tr A $= a - b + c + 4$

From the definition,

(DB.7) Tr **(AB)** $= A_{ii}B_{jj} = B_{jj}A_{ii} = $ Tr **(BA)**

(DB.8) The determinant of **A**, written as det A, is defined as:

(a) for a two by two matrix,

$$A = \begin{pmatrix} a & b \\ c & d \end{pmatrix} \rightarrow \det A = \begin{vmatrix} a & b \\ c & d \end{vmatrix} = ad - bc$$

(b) for a three by three matrix,

$$A = \begin{pmatrix} a & b & c \\ d & e & f \\ g & h & i \end{pmatrix}$$

$$\rightarrow \det A = a \begin{vmatrix} e & f \\ h & i \end{vmatrix} - b \begin{vmatrix} d & f \\ g & i \end{vmatrix} + c \begin{vmatrix} d & e \\ g & h \end{vmatrix}$$

$$= a(ei - fh) - b(di - fg) + c(dh - eg)$$

DC. Four-Vector Formalism Product

The dot product is defined as,

(DC.1) $A \cdot B = A_\mu B^\mu = g^{\mu\nu} A_\mu B_\nu = g_{\mu\nu} A^\nu B^\mu$

Where

(DC.2) $g^{\mu\nu} = g_{\mu\nu} = \begin{pmatrix} 1 & 0 & 0 & 0 \\ 0 & -1 & 0 & 0 \\ 0 & 0 & -1 & 0 \\ 0 & 0 & 0 & -1 \end{pmatrix}$

Carry on the multiplication by separating temporal components from the spatial, we get

$$\rightarrow A \cdot B = g_{00} A_0 B_0 + g_{ij} A_i B_j$$
$$= A_0 B_0 - A_i B_i \text{ where } i = 1, 2, 3$$

Appendix E

Calculus

EA. Derivatives

The derivative is generally defined for a continuous function as:

(EA.1) $\dfrac{df(x)}{dx} = \lim\limits_{\Delta x \to 0} \dfrac{f(x+\Delta x)-f(x)}{\Delta x}$

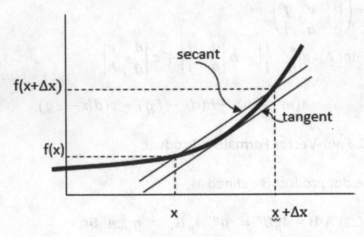

Fig.EA.1

Note that as $\Delta x \to 0$, the secant line approaches the tangent line at one point on the curve defined by f(x).

Nomenclature: To differentiate is to take derivatives.

Here are some standard derivatives:

(EA.2) $\dfrac{d}{dx}$(constant) = 0

(EA.3) $\dfrac{dx^n}{dx} = nx^{n-1}$

(EA.4) $\dfrac{d\sin x}{dx} = \cos x$

(EA.5) $\dfrac{d\cos x}{dx} = -\sin x$

(EA.6) $\dfrac{de^{kx}}{dx} = k\,e^{kx}$

(EA.7) $\dfrac{d\ln(x)}{dx} = \dfrac{1}{x}$

(EA.8) $\dfrac{d(\log_a(x))}{dx} = \dfrac{1}{x\ln(a)}$

Chain rule

(EA.9a) $\dfrac{dy}{dt} = \dfrac{dy}{dx}\dfrac{dx}{dt}$

Product rule or Leibniz rule

(EA.9b) $d(uv) = u\,dv + v\,du$

Some identities for the del operator $\nabla \equiv \vec{\imath}\dfrac{\partial}{\partial x} + \vec{\jmath}\dfrac{\partial}{\partial y} + \vec{k}\dfrac{\partial}{\partial z}$:

(EA.10) $\nabla(fg) = f\nabla g + g\nabla f$

(EA.11) $\nabla f\vec{v} = f\nabla \cdot \vec{v} + \vec{v} \cdot \nabla f$

(EA.12) $\nabla(\vec{u}\, x\, \vec{v}) = \vec{v} \cdot (\nabla\, x\, \vec{u}) - \vec{u} \cdot (\nabla\, x\, \vec{v})$

(EA.13) $\nabla\, x\, (\nabla f) = 0$

(EA.14) $\nabla \cdot (\nabla\, x\, \vec{u}) = 0$

(EA.15) $\nabla \cdot (\nabla f) = \nabla^2 f$

(EA.16) In 4-d, $\partial^2 \equiv \partial_t^2 - \nabla^2$, where $\partial_t \equiv \frac{\partial}{\partial t}$

Partial Derivatives

If a function has several arguments, then we can take the partial derivative with respect to any of the arguments, one at a time.

Suppose that $f \rightarrow f(u, v)$ then a partial derivative is:

(EA.17) $f_u = \frac{\partial f(u,v)}{\partial u}$ or $f_v = \frac{\partial f(u,v)}{\partial v}$

The total derivative is then:

(EA.18) $df(u, v) = f_u du + f_v dv$

$$= \frac{\partial f(u,v)}{\partial u} du + \frac{\partial f(u,v)}{\partial v} dv$$

This can be extended to any number of parameters.

Legendre transform

Consider a function of two variables, f(x,y). Its differential is,

(EA.19) $df = \frac{\partial f}{\partial x} dx + \frac{\partial f}{\partial y} dy$

Define $\frac{\partial f}{\partial x} \equiv u$ and $\frac{\partial f}{\partial y} \equiv v$

So equation EA.19 can now be written as,

(EA.20) $df = u dx + v dy$

So x and u are conjugate pairs, so with y and v. Consider the vy and compute the differential using the product rule (EA.9),

(EA.21) $d(vy) = vdy + ydv$

Subtract this from EA.20,

(EA.22) $df - d(vy) = udx + vdy - vdy - ydv$

$$d(f - vy) = udx - ydv$$

$$dg = udx - ydv$$

Where $g \equiv f - vy$ is the Legendre transform. What we have done is take a function $f(x, y)$ and transforms it to a function $g(x, v)$, where y and v are conjugate pairs.

Taylor Expansion

The expansion of a function $f(x)$ about a point x = a, is given by

(EA.23)

$$f(x) = f(a) + f^{(1)}(a)(x - a) + \frac{1}{2!}f^{(2)}(a)(x - a)^2$$

$$+ \frac{1}{3!}f^{(3)}(a)(x - a)^3 + \cdots + \frac{1}{n!}f^{(n)}(a)(x - a)^n$$

Where $f^{(n)}$ denotes the nth derivative of $f(x)$ evaluated at point a.

EB. Integrals

Definition

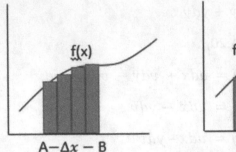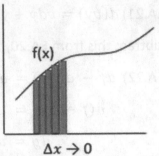

$$A-\Delta x - B \qquad\qquad \Delta x \to 0$$

Fig. EB.1

(EB.1) The integral $\int_A^B f(x)$ is the area under the curve drawn by the function f(x) and the x-axis between points A and B. We draw slices of Δx. As we let $\Delta x \to 0$, the slices will occupy the exact area under the curve.

Note that integration is the opposite operator of the derivative:

(EB.2) $d(\int(f(x))) = \int d\,(f(x)) = f(x)$

This follows from the Fundamental Theorem of Calculus:

(EB.3) $\int_a^b f(x)dx = F(b) - F(a)$

Where $f(x) = F'(x) \equiv \dfrac{dF(x)}{dx}$

Integration by parts

(EB.4) $\int$ udv = uv $- \int$ v du

When uv = 0, $\int$ udv $= - \int$ v du

Some important integrals:

(EB.5) $\int x^n \, dx = \dfrac{x^{n+1}}{n+1}$

(EB.6) $\int \sin x \, dx = -\cos x$

(EB.7) $\int \cos x \, dx = \sin x$

(EB.8) $\int \sin^2 x \, dx = \int \frac{1}{2}(1 - \cos 2x)dx = \frac{x}{2} - \sin x$

(EB.9) $\int e^x \, dx = e^x$

EC. Gauss Theorem

The theorem relates the integral of a derivative of a function ($\nabla \cdot F$) over a region to the integral of the original function F over the boundary of that region.

Let V be a simple solid region and S the boundary surface of V. Let F be a vector field with components having continuous partial derivatives in V. Then

$$\text{EC.1} \quad \iint_S F \cdot dS = \iiint_V \nabla \cdot F \, dV$$

Note: this allows to go from a triple integral to a double integral, and vice-versa.

ED. The Jacobian

We want to know how to change variables in an integral. In general, we consider a change of variables as a transformation T from a uv-plane to an xy-plane.

(ED.1) $T(u, v) = (x, y)$

(ED.2) $x = g(u, v)$ and $y = h(u, v)$

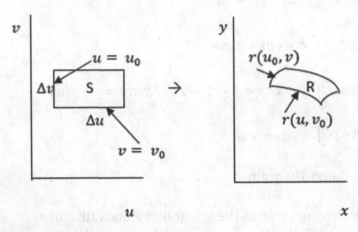

Fig. ED.1

We start with a small rectangle in the uv-plane, whose dimensions are Δu by Δv (Fig. ED.1). The image of S in the uv-plane is some region R in the xy-plane. The vector in the xy-plane

(ED.3) $r = xi + yj = g(u, v)\, i + h(u, v)\, j$

is the position vector of the image of the point (u, v). We approximate the image R as a parallelogram determined by secant vectors (Fig. ED.2)

(ED.4) $a = r(u_0 + \Delta u, v_0) - r(u_0, v_0)$

and $b = r(u_0, v_0 + \Delta v) - r(u_0, v_0)$

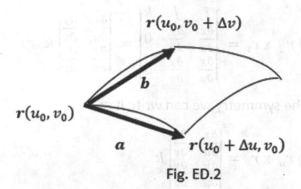

Fig. ED.2

Using the definition of the derivative (equation EA.1), we have

(ED.5a) $r_u = \lim\limits_{\Delta u \to 0} \dfrac{r(u_0+\Delta u, v_0) - r(u_0, v_0)}{\Delta u}$

Or

(ED.5b) $a = r_u \Delta u$

Similarly,

(ED.6a) $r_v = \lim\limits_{\Delta v \to 0} \dfrac{(u_0, v_0+\Delta v) - r(u_0, v_0)}{\Delta v}$

Or

(ED.6b) $b = r_v \Delta v$

The area = length by width,

(ED.7) $|a \times b| = |r_u \Delta u \times r_v \Delta v| = |r_u \times r_v| \Delta u \Delta v$

Computing the cross product, (equation DA.4a)

$$\text{(ED.8a) } r_u \times r_v = \begin{vmatrix} i & j & k \\ \frac{\partial x}{\partial u} & \frac{\partial y}{\partial u} & 0 \\ \frac{\partial x}{\partial v} & \frac{\partial y}{\partial v} & 0 \end{vmatrix} = \begin{vmatrix} \frac{\partial x}{\partial u} & \frac{\partial y}{\partial u} \\ \frac{\partial x}{\partial v} & \frac{\partial y}{\partial v} \end{vmatrix} k$$

Using the symmetry we can write it as,

$$\text{(ED.8a) } r_u \times r_v = \begin{vmatrix} \frac{\partial x}{\partial u} & \frac{\partial x}{\partial v} \\ \frac{\partial y}{\partial u} & \frac{\partial y}{\partial v} \end{vmatrix} k$$

The determinant in this calculation is known as the Jacobian. And it is designated by,

$$\text{(ED.9) } \frac{\partial(x,y)}{\partial(u,v)} \equiv \begin{vmatrix} \frac{\partial x}{\partial u} & \frac{\partial x}{\partial v} \\ \frac{\partial y}{\partial u} & \frac{\partial y}{\partial v} \end{vmatrix} = \frac{\partial x}{\partial u}\frac{\partial y}{\partial v} - \frac{\partial x}{\partial v}\frac{\partial y}{\partial u}$$

It is how the area is transformed.

$$\text{(ED.10) } \Delta A = \left| \frac{\partial(x,y)}{\partial(u,v)} \right| \Delta u \Delta v$$

Hence, a double integral which is itself an area under a curve will transforms as,

$$\text{(ED.11) } \iint f(x,y)\ dxdy = \iint f(x,y)\ dA$$

$$= \iint f(g(u,v), h(u,v))\ \left| \frac{\partial(x,y)}{\partial(u,v)} \right| dudv$$

Example: polar coordinates are given by (CB.1-2)

(ED.12a) x = r cos θ

(ED.12b) y = r sin θ

The Jacobian is

$$\text{(ED.13)} \quad \frac{\partial(x,y)}{\partial(r,\theta)} = \begin{vmatrix} \frac{\partial x}{\partial r} & \frac{\partial x}{\partial \theta} \\ \frac{\partial y}{\partial r} & \frac{\partial y}{\partial \theta} \end{vmatrix} = \begin{vmatrix} \cos\theta & -r\sin\theta \\ \sin\theta & r\cos\theta \end{vmatrix}$$

$$= r\cos^2\theta + r\sin^2\theta = r$$

The area is

(ED.14) $\Delta x \Delta x = r \Delta r \Delta \theta$

The integral transforms as,

(ED.15) $\iint f(x,y) \, dxdy = \iint f(r\cos\theta, \sin\theta) \, rdrd\theta$

Appendix F

FA. Lagrange Multipliers

We need to know how to maximize (minimize) a quantity in terms of its arguments that are subject to certain constraints. For example suppose the quantity in question is $x^2 + y^2$ subject to the constraint that $x + 2y - 1 = 0$ (Fig. FA.1).

Solving for x in terms of y,

→ $x = 1 - 2y$

→ Let $F(x,y) = x^2 + y^2$

→Then $F(x,y) = (1 - 2y)^2 + y^2$

$$= 1 - 4y + 5y^2$$

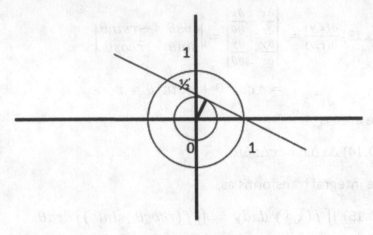

Fig. FA.1

To find the minimum, we take the derivative with respect to y, and let it equal to zero,

$$\rightarrow \frac{\partial F}{\partial y} = -4 + 10y = 0$$

$$\rightarrow \text{or } y = \frac{2}{5}$$

$$\rightarrow \text{Therefore, } x = 1-2y = \frac{1}{5}$$

In other words, we were looking for the small distance from the origin that satisfied both $x^2 + y^2$ and $x + 2y - 1 = 0$. But what if the functions are more complicated with many more constraints, the solution might not be found easily.

We restate the problem as such: find the maximum (minimum) for a function F(x,y) given that G(x,y) = 0.

We multiply the constraint by a constant λ, called the Lagrangian multiplier, and add that to the function F.

$$F(x,y) + \lambda\, G(x,y) = 0$$

We differentiate this equation first with respect to x and then with respect to y.

$$\frac{\partial F}{\partial x} + \lambda \frac{\partial G}{\partial x} = 0$$

$$\frac{\partial F}{\partial y} + \lambda \frac{\partial G}{\partial y} = 0$$

We now have two equations and we can solve for each of the variables, x and y, in terms of λ.

From the previous example,

$$F(x,y) = x^2 + y^2 , \; G(x,y) = x + 2y - 1 = 0$$

→ $2x + \lambda = 0$

→ $2y + 2\lambda = 0$

Solving for x and y,

→ $x = -\lambda/2$

→ $y = -\lambda$

Substitute those results into the constraint,

→ $x + 2y - 1 = 0$

→ $(-\lambda/2) + 2(-\lambda) - 1 = 0$

$$\rightarrow -\frac{5}{2}\lambda = 1$$

$$\rightarrow \lambda = -\frac{2}{5}$$

And we get the same results for x and y.

This method is easier when we have a larger number of variables, say n, subject to a number of constraints, say m, with m ≤ n. For a function $F(x_1, x_2, ... x_n)$, we introduce a multiplier for each of the constraints, that is, $\lambda_1 \lambda_2 ... \lambda_m$ and follow the procedure in the exact way as we've outlined so far.

Appendix G

The Stirling Approximation Formula

$$\ln N! \approx N \ln N - N$$

Proof:

Consider (i) $\frac{d}{dx}(\ln(x)) = \frac{1}{x}$ and $\frac{d}{dx}(x) = 1$

(ii) $\frac{d}{dx}(x\ln(x)) = \ln(x) + 1$

(iii) $\frac{d}{dx}(x\ln(x) - x) = \ln(x)$

Take the integral on both sides.

(iv) $\int \frac{d}{dx}(x\ln(x) - x)\,dx = \int \ln(x)\,dx$

(v) $x\ln(x) - x = \int \ln(x)\,dx$

Putting the limits of integration:

(vi) $\int_1^n \ln(x)\, dx = [x\ln(x) - x]_1^n$

$$= n\ln(n) - n - (\ln(1) - 1)$$

For very large n, we can ignore the last two terms.

(vii) $\int_1^n \ln(x)\, dx \approx n\ln(n) - n$

To evaluate the LHS, consider the following graph,

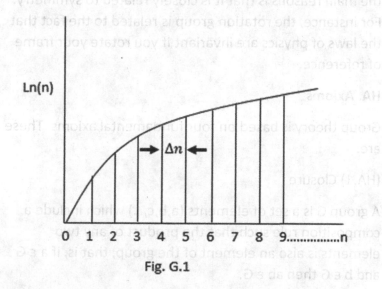

Fig. G.1

The LHS is just the total area under the curve, which we can now evaluate as,

(viii) Area $= \Delta n\ln(1) + \Delta n\ln(2) + \Delta n\ln(3) + \cdots + \Delta n\ln(n)$

$$= \Delta n(\ln(1) + \ln(2) + \ln(3) + \cdots + \ln(n))$$

$$= \Delta n(\ln(1 \cdot 2 \cdot 3 \cdot \ldots \cdot n))$$

$$= \Delta n(\ln(n!))$$

But since $\Delta n = 1$

Therefore, $\int_1^n \ln(x)\, dx = \ln(n!) \approx n\ln(n) - n$

Appendix H

Group Theory

Group theory plays a fundamental role in physics. One of the main reasons is that it is closely related to symmetry. For instance, the rotation group is related to the fact that the laws of physics are invariant if you rotate your frame of reference.

HA. Axioms

Group theory is based on four fundamental axioms. These are:

(HA.1) Closure

A group G is a set of elements (a,b,c,...) which include a composition rule such that the product of any two elements is also an element of the group; that is, if a ε G and b ε G then ab ε G.

(HA.2) Associativity

The composition rule is associative, meaning (ab)c = a(bc)

From axiom (i), a,b,c ε G

Suppose the LHS is: ab = p, and (ab)c = pc = t, where p,t ε G

For the RHS: bc= q, and a(bc) = aq, where q ε G

The associative rule demands that LHS = RHS, and therefore t = aq.

(HA.3) Identity

The group G has an element e, called the identity element, such that for every element a, ae = ea = a. Note that the identity element for the group is unique.

(HA.4) Inverse

For every element a ε G there exists an inverse, denoted by a^{-1} such that a a^{-1} = $a^{-1}a$ = e, where again e is the identity element.

HB. Representation of Groups

A representation is a mapping that takes a group element g ε G into a linear operator F, denoted by,

$$F: g \rightarrow F(G)$$

Such that the composition rule is preserved, meaning:

(HB.1) Closure: F(a)F(b) = F(ab)

(HB.2) The identity is preserved: F(e) = e

The groups G and F(G) are said to be isomorphic, meaning that both groups have the same mathematical structure.

HC. The Rotation Group

A function can be viewed as given an input x, and the function uses that input to yield an output y:

$$f : x \rightarrow y$$

Or, what is commonly known as y = f(x). Similarly, the individual elements of a group are outputs of function, and the inputs are called parameters.

The rotation group is the set of all rotations about the origin. In 2-D, the elements contain one-parameter elements obeying the composition rule,

(HC.1) $R(\theta_1)R(\theta_2) = R(\theta_1 + \theta_2)$

The inverse is,

(HC.2) $R^{-1}(\theta) = R(-\theta)$

And the identity element is,

(HC.3) $I = R(0)$

Representation of the rotation group: consider a vector with coordinates x_i. Let x_i' be the coordinates of the vector rotated by an angle θ in the plane. The components of these two vectors are related by the transformation:

(HC.4) $x_i' = R_{ij}x_i$

In this case, the rotation can be represented by a 2-D matrix,

(HC.5) $R(\theta) = \begin{pmatrix} \cos\theta & \sin\theta \\ -\sin\theta & \cos\theta \end{pmatrix}$

Consider the transpose of this matrix,

(HC.6) $R^T(\theta) = \begin{pmatrix} \cos\theta & -\sin\theta \\ \sin\theta & \cos\theta \end{pmatrix}$

Multiply the above matrices,

(HC.7)

$\rightarrow R(\theta)R^T(\theta) = \begin{pmatrix} \cos\theta & \sin\theta \\ -\sin\theta & \cos\theta \end{pmatrix}\begin{pmatrix} \cos\theta & -\sin\theta \\ \sin\theta & \cos\theta \end{pmatrix}$

$$= \begin{pmatrix} \cos^2\theta + \sin^2\theta & -\cos\theta\sin\theta + \cos\theta\sin\theta \\ -\sin\theta\cos\theta + \cos\theta\sin\theta & \cos^2\theta + \sin^2\theta \end{pmatrix}$$

$$= \begin{pmatrix} 1 & 0 \\ 0 & 1 \end{pmatrix} = I$$

This means that the transpose is also the inverse,

(HC.8) $R^{-1}(\theta) = R^T(\theta)$

Another important feature is the determinant,

(HC.9) $det\ R(\theta) = det \begin{pmatrix} \cos\theta & \sin\theta \\ -\sin\theta & \cos\theta \end{pmatrix}$

$$= \cos^2\theta + \sin^2\theta = 1$$

The 2-D rotation group is part of a larger group, the group $SO(N)$, which are special orthogonal $N\ X\ N$ matrices. They are orthogonal because $R^{-1}(\theta) = R^T(\theta)$ (equation HC.8), and special because $det\ R(\theta) = 1$ (equation HC.9).

This treatment can be extended to higher dimensions. For example in 3-D, we would need a set of two-parameter elements: $R \rightarrow R(\theta_1, \theta_2)$.

Note that the parameter θ is continuous and so the group SO(N) is said to be a continuous group.

The range of the parameter θ is between 0... 2π. We say that the group SO(N) is compact.

If every elements of the group commutes among themselves, that is, for any elements a ε G and b ε G, we have ab = ba, the group is said to be abelian. If not, the group is said to be non-abelian.

If the number of elements in the group G is finite, the group is said to be finite. Likewise, if the number of elements is infinite, the group is infinite.

HD. Lie Groups and Generators

Here we confine ourselves to a group G that has the following properties:

(i) There is a finite set of continuous parameters θ_i.

(ii) There exists derivatives of the group elements with respect to all the parameters.

We call this group a Lie group.

One of the key features of Lie groups is the notion of generators. To illustrate this, we will consider a one-parameter element of a Lie group G.

We obtain the identity element by setting $\theta = 0$,

(HD.1) $g(\theta)\big|_{\theta=0} = e$

The generators of the Lie group are defined as the derivatives of the group element with respect to the parameter at $\theta = 0$:

(HD.2) $X = \dfrac{\partial g}{\partial \theta}\bigg|_{\theta=0}$

Where X is the generator. If there are n parameters in the Lie group G, then we have n generators:

(HD.3) $X_i = \dfrac{\partial g}{\partial \theta_i}\bigg|_{\theta=0}$ where $i = 1 \dots n$

Because Hermitian operators $(X^T = X)$ in QM are ubiquitous, we want a unitary representation (see next section) and choose the generators to be Hermitian, that is,

(HD.4) $X_i = -i \frac{\partial g}{\partial \theta_i}\Big|_{\theta=0}$

This unitary representation is best expressed by considering some changes in the parameter θ_i around the origin. We denote the representation by D and express it in terms of the exponential:

(HD.5) $D(\theta) = e^{i\theta X}$

We can check that it is unitary:

(HD.6) $D^\dagger(\theta)D(\theta) = e^{-i\theta X}e^{i\theta X} = 1$

The importance of the generators is that they themselves form a vector space and obey the commutation relation:

(HD.7) $[X_i, X_j] = if_{ijk}X_k$

Where f_{ijk} are the structure constants, and the above equation forms the composition rule of a Lie algebra.

HE. Unitary Groups

Complex numbers are ubiquitous in QM, and to deal with that reality, we need the unitary representations as discussed above. The unitary group U(n) consists of $N \times N$ unitary matrices, for which we have,

(HE.1) $U^\dagger U = 1$ or $U^\dagger = U^{-1}$

The special unitary group is denoted by SU(N), for which det = 1.

The number of generators for a U(N) group is N^2, and for a SU(N) group, $N^2 - 1$.

The simplest unitary group is the group U(1), which is a 1×1 matrix, that is, the complex number,

(HE.2) $U(\theta) = e^{-i\theta}$

This is the familiar unit circle. We have one generator, $(N^2 = 1^2 = 1)$.

The next unitary group of interest is the SU(2) group with 3 generators $(N^2 - 1 = 2^2 - 1 = 3)$. These are related to the Pauli matrices, with the following commutation relationship,

(HE.3) $\left[\frac{\sigma_i}{2}, \frac{\sigma_j}{2}\right] = i\varepsilon_{ijk}\frac{\sigma_k}{2}$

And the Pauli matrices are,

(HE.4) $\sigma_1 = \begin{pmatrix} 0 & 1 \\ 1 & 0 \end{pmatrix}, \sigma_2 = \begin{pmatrix} 0 & -i \\ i & 0 \end{pmatrix}, \sigma_3 = \begin{pmatrix} 1 & 0 \\ 0 & -1 \end{pmatrix}$

We can write an element of SU(2) as,

(HE.5) $U = e^{i\alpha_i\sigma_i/2}$

With α_i being a real number.

Of great interest is the case of the weak nuclear force which is the product of the groups U(1) X SU(2). Consider a Hermitian matrix of the form,

(HE.6) $H = \begin{pmatrix} \alpha_0 + \alpha_3 & \alpha_1 - i\alpha_2 \\ \alpha_1 + i\alpha_2 & \alpha_0 - \alpha_3 \end{pmatrix}$

We can now split this matrix as,

(HE.7) $H = \alpha_0 I + \alpha_k \sigma_k$

Where k now runs from 1 to 3.

A general member of this group is,

(HE.8) $U = \exp(\alpha_0 I + \alpha_k \sigma_k) = \exp(\alpha_0 I) \exp(\alpha_k \sigma_k)$

The phase factor $\exp(\alpha_0 I)$ belong to a U(1) group (equation HE.2), and the second phase factor $\exp(\alpha_k \sigma_k)$ to SU(2) (equation HE.5). This element belongs to the product U(1) X SU(2).

Appendix I

Gaussian Integrals

IA.

(IA.1) $C = \int e^{-y^2} dy$, integral is from $-\infty$ to $+\infty$.

Squaring both sides,

(IA.2) $C^2 = \int e^{-y^2} dy \int e^{-x^2} dx$

In the second term, we've replaced y by x, since these are just dummy variable in the integration,

(IA.3) $C^2 = \int\int e^{-(y^2 + x^2)} dxdy$

Switching to polar coordinates (equationsCB.1-2),

(IA.4) Let $x = r \cos\theta$

(IA.5) And $y = r \sin\theta$

(IA.6) Then $y^2 + x^2 = (r \cos\theta)^2 + (r \sin\theta)^2$

$$= r^2 \cos^2\theta + r^2 \sin^2\theta$$
$$= r^2 (\cos^2\theta + \sin^2\theta)$$
$$= r^2$$

(IA.7) Therefore, $C^2 = \iint e^{-r^2} dxdy$

Also, the product dydx is just an element of the area of a small square. In polar coordinates, that area is rdrdθ (equation ED.15). So,

(IA.8) $C^2 = \int_0^\infty e^{-r^2} rdr \int_0^{2\pi} d\theta$

Make another change in variable, $u = r^2$, so that $du = 2rdr$

(IA.9) $C^2 = \frac{1}{2} \int_0^\infty e^{-u} du \int_0^{2\pi} d\theta$

$$= \frac{1}{2} (-) (e^{-\infty} - e^{-0})(2\pi - 0)$$

$$= \frac{1}{2} (-) (0 - 1)(2\pi)$$

$$= \pi$$

(IA.10) Therefore, $C = \int_{-\infty}^{+\infty} e^{-y^2} dy = \pi^{\frac{1}{2}}$

IB.

(IB.1) $C = \int_{-\infty}^{+\infty} e^{-ay^2} dy$

In the case that a constant "a" multiplies y^2, we make the following substitution,

(IB.2) $x^2 = ay^2$

(IB.3)Then $x = a^{\frac{1}{2}}y$

(IB.4) Taking derivatives,

$dx = a^{\frac{1}{2}}dy \rightarrow dy = a^{-\frac{1}{2}}dx$

Substituting IB.4 into IB.1,

(IB.5) $C = \int_{-\infty}^{+\infty} e^{-ay^2} dy = a^{-\frac{1}{2}}\int_{-\infty}^{+\infty} e^{-x^2} dx$

$\quad\quad = a^{-\frac{1}{2}} \pi^{\frac{1}{2}}$, from AI.10

$\quad\quad = (\pi/a)^{\frac{1}{2}}$

IC.

(IC.1) $C = \int_{-\infty}^{+\infty} e^{-ay^2 + by} dy$

(IC.2) The exponent is

$-ay^2 + by = - a[y^2 - (b/a)y]$.

(IC. 3) We complete the square on the RHS:
$$= -a[y^2 - (b/a)y + (b/2a)^2 - (b/2a)^2]$$
$$= -a[(y - b/2a)^2 - (b^2/4a^2]$$
$$= -a(y - b/2a)^2 + b^2/4a$$

(IC.4) Let $x^2 = (y - b/2a)^2$

(IC.5) $x = (y - b/2a)$

(IC.6) $dx = dy$

(IC.7) $C = \int_{-\infty}^{+\infty} e^{-ay^2+by} dy$, from IC.1

$$= \int_{-\infty}^{+\infty} e^{-a\left(y-\frac{b}{2a}\right)^2 + \frac{b^2}{4a}} dy \text{ , from IC.3}$$

$$= e^{\frac{b^2}{4a}} \int_{-\infty}^{+\infty} e^{-ax^2} dx \text{ , from IC.4, IC.6}$$

$$= (\pi/a)^{\frac{1}{2}} e^{\frac{b^2}{4a}} \text{ , from IB.5}$$

ID.

We will borrow from the previous sections:

(ID.1) $C_1 = \int_{-\infty}^{\infty} e^{-\frac{1}{2}y^2} dy = (2\pi)^{\frac{1}{2}}$ (IA.10)

(ID.2) $C_2 = \int_{-\infty}^{\infty} e^{-\frac{1}{2}ay^2} dy = (2\pi/a)^{\frac{1}{2}}$ (IB.5)

(ID.3) $C_3 = \int_{-\infty}^{\infty} e^{-\frac{1}{2}ay^2+by} dy = (2\pi/a)^{\frac{1}{2}} e^{b^2/2a}$ (IC.7)

We want to generalize this for an n X n matrix A, we rewrite the integral as,

(ID.4) $C_2 \rightarrow C_4$

$$= \int_{-\infty}^{\infty} \int_{-\infty}^{\infty} \dots \int_{-\infty}^{\infty} dx_1 dx_2 \dots dx_n \, e^{-\frac{1}{2}x \cdot A \cdot x}$$

(ID.5) $C_3 \rightarrow C_5$

$$= \int_{-\infty}^{\infty} \int_{-\infty}^{\infty} \dots \int_{-\infty}^{\infty} dx_1 dx_2 \dots dx_n \, e^{-\frac{1}{2}x \cdot A \cdot x + J \cdot x}$$

where $x \cdot A \cdot x = x_i A_{ij} x_j$ and $J \cdot x = J_i x_i$, with $I,j = 1,2...N$, and repeated indices summed over.

We will calculate for N=2, and then generalize to any N. We take any 2x2 matrix A' and diagonalize it to A.

(ID.6)

$$A' = \begin{bmatrix} a & b \\ c & d \end{bmatrix} \rightarrow A = \begin{bmatrix} ad-bc & 0 \\ 0 & 1 \end{bmatrix}$$

Note: det A' = det A = ad − bc

We calculate

(ID.7) $\rightarrow x_i A_{ij} x_j = x_1(A_{1j}x_j) + x_2(A_{2j}x_j)$, i=1,2

$$= x_1(A_{11}x_1 + A_{12}x_2) + x_2(A_{21}x_1 + A_{22}x_2), \text{j=1,2}$$

But $A_{12} = A_{21} = 0$

Therefore,

$\rightarrow x_i A_{ij} x_j = x_1 A_{11} x_1 + x_2 A_{22} x_2$

$$= (ad - bc)(x_1)^2 + (x_2)^2$$

$$= (Det[A])(x_1)^2 + (x_2)^2$$

(ID.8) So we take $C_4 = = \int_{-\infty}^{\infty} \int_{-\infty}^{\infty} dx_1 dx_2 e^{-\frac{1}{2}x \cdot A \cdot x}$, N=2

This becomes,

(ID.9) $C_4 = \int_{-\infty}^{\infty} \int_{-\infty}^{\infty} dx_1 dx_2 e^{-\frac{1}{2}(Det[A](x_1)^2 + (x_2)^2)}$

$$= \int_{-\infty}^{\infty} dx_2 e^{-\frac{1}{2}(x_2)^2} \int_{-\infty}^{\infty} dx_1 e^{-\frac{1}{2}Det[A](x_1)^2}$$

The first integral is C_1, and the second is C_2.

(ID.10) $C_4 = (2\pi)^{\frac{1}{2}}(2\pi/det[A])^{\frac{1}{2}} = ((2\pi)^2/det[A])^{\frac{1}{2}}$

For any nxn matrix A,

(ID.11) $C_4 = ((2\pi)^N/det[A])^{\frac{1}{2}}$

(ID.12) $C_5 = \int_{-\infty}^{+\infty} \int_{-\infty}^{+\infty} \int_{-\infty}^{+\infty} dx_1 dx_2 dx_3 \dots dx_n \; e^{-\frac{1}{2}x \cdot A \cdot x + J \cdot x}$

$$= ((2\pi)^N/det[A])^{\frac{1}{2}} \; e^{\frac{1}{2}J \cdot A^{-1} \cdot J}$$

IE.

By differentiating IB.5 with respect to a, we get all the integrals of the form:

(IE.1) $C_{2n}(a) = \int_{-\infty}^{\infty} y^{2n} e^{-\frac{1}{2}ay^2} dy$

(IE.2) Example: $C_2(a) = \int_{-\infty}^{\infty} y^2 e^{-\frac{1}{2}ay^2} dy$

Take the derivative with respect to a,

$$\rightarrow \int_{-\infty}^{\infty} y^2 e^{-\frac{1}{2}ay^2} dy = -\frac{\partial}{\partial a} \int_{-\infty}^{\infty} e^{-\frac{1}{2}ay^2} dy$$

$$= -\frac{\partial}{\partial a}(\pi/a)^{\frac{1}{2}} = -\frac{\partial}{\partial a}(\pi^{\frac{1}{2}} a^{-\frac{1}{2}})$$

392

$$= -\pi^{\frac{1}{2}}(-\frac{1}{2}\,a^{-3/2}) = \frac{1}{2}\sqrt{\frac{\pi}{a^3}}$$

(IE.3) Example: $C_4(a) = \int_{-\infty}^{\infty} y^4 e^{-\frac{1}{2}ay^2}\,dy$

We need to take the derivative with respect to a twice.

$\rightarrow \int_{-\infty}^{\infty} y^4 e^{-\frac{1}{2}ay^2}\,dy = (-\frac{\partial}{\partial a})(-\frac{\partial}{\partial a})\int_{-\infty}^{\infty} e^{-\frac{1}{2}ay^2}\,dy$

$= (-\frac{\partial}{\partial a})\,(\frac{1}{2}\sqrt{\frac{\pi}{a^3}})$ (from equation IE.2)

$= \frac{3}{4}\sqrt{\frac{\pi}{a^5}}$

Appendix J

Special Functions

JA. Fourier Transform

If $f(x)$ is periodic with a period L, then

(JA.1) $f(x) = \sum_{n=-\infty}^{n=\infty} a_n e^{ik_n x}$

Where

(JA.2) $k_n = \frac{2\pi n}{L}$

The coefficients a_n are given by

(JA.3) $a_n = \frac{1}{L}\int_0^L f(x)e^{-ik_n x}dx$

The Fourier transform of the function f(x) is defined as:

(JA.4) $F(k) = \int_{-\infty}^{+\infty} f(x)e^{-ikx}dx$

Where $f(x)$ is the Fourier component of $F(k)$.

The inverse Fourier transform is then,

(JA.5) $f(x) = \frac{1}{\sqrt{2\pi}} \int_{-\infty}^{+\infty} F(k)e^{ikx}dx$

JB. Dirac Delta Function

The Dirac delta-function is defined as

(JB.1) $\int_{-\infty}^{+\infty} f(x)\delta(x - x_0)dx = f(x_0)$

Some important identities:

(JB.2) $\delta(-x) = \delta(x)$

(JB.3) $\delta(cx) = \frac{1}{c}\delta(x)$ for a constant $c > 0$

(JB.4) $\delta[f(x)] = \sum_i \frac{1}{f'(x_i)}\delta(x - x_i)$

Where $f'(x_i)$ is the derivative of $f(x)$ wrt x_i.

Proof: we make a Taylor expansion (appendix E, equation EA.23) of the function $f(x)$:

$\rightarrow f(x) = f(x_i) + (x - x_i)(\frac{df}{dx})_{x=x_i} + \cdots$

The δ-function has non-zero contributions for each of the roots x_i of the form:

$\rightarrow \delta[f(x)] = \sum_i \delta[(x - x_i)(\frac{df}{dx})_{x=x_i}]$

Using JB.3, we have $\delta[f(x)] = \sum_i \frac{1}{f'(x_i)}\delta(x - x_i)$

(JB.5) $\delta(x - y) = \frac{1}{2\pi} \int_{-\infty}^{+\infty} e^{ip(x-y)} dp$

JC. Heaviside Function

The Heaviside function is defined as:

(JC.1) $\theta(x) = 1 \quad for \ x > 0$

$\qquad\qquad = 0 \quad for \ x < 0$

The relation between the Dirac delta function and the Heaviside function is,

(JC.2) $\frac{\partial}{\partial t_x} \theta(t_x - t_y) = \delta(t_x - t_y)$

Appendix K

Calculus of Complex Functions

KA. Complex Numbers

A complex number can be designated as:

(KA.1) z = x + iy

Where x is a real number that forms the real part (x = RE z) of the complex number z, and y is also a real number but forms the imaginary part (y = Im z), and i is the imaginary number ($i^2 = -1, or \ i = \sqrt{-1}$).

The complex conjugate of z is defined as:

(KA.2) z* = x - iy

KB. Geometric representation

The complex plane (also known as the Argand diagram) is represented with x on the real axis, y on the imaginary axis. Then z can be denoted as an ordered pair:

(KB.1) z = (x,y)

Note in Fig. KB.1, with x = 0, and y = 1 in KA.1, the imaginary number is the point $i = (0, 1)$

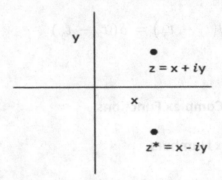

Fig. KB.1

KC. Differentiability

Consider that the complex conjugate z* is not differentiable. Here's why. The derivative of any function is defined as (equation EA.1):

(KC.1) $\frac{df(z)}{dz} = \lim_{\Delta z \to 0} \frac{f(z+\Delta z)-f(z)}{\Delta z}$

In the case we're interested, we have,

(KC.2) $f(z) = z^* = x - iy$

Now we write (from equation KA.1),

(KC.3) $\Delta z = \Delta x + i\,\Delta y$

Hence we have,

(KC.3) $\dfrac{f(z+\Delta z)-f(z)}{\Delta z} = \dfrac{(z+\Delta z)^* - z^*}{\Delta z}$

$$= \dfrac{\Delta z^*}{\Delta z} = \dfrac{\Delta x - i\Delta y}{\Delta x + \Delta y}$$

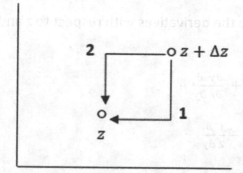

Fig. KC. 1

As $\Delta z \to 0$, we consider two paths as indicated in Fig. KC.1, where we first take $\Delta y = 0$ (path 1), the derivative is +1; while for path 2, $\Delta x = 0$, the derivative is now -1. So the limit $\Delta z \to 0$ does not exist. Consequently any function that contains z^* will not be differentiable.

We define a function f(z) to be holomorphic if it is differentiable at all points in a domain D.

KD. Cauchy-Riemann Equations

First we will express x and y in terms of z and z*:

(KD.1) $x = \frac{z+z^*}{2}$ and $y = \frac{z-z^*}{2i}$

Next we take the derivatives with respect to z and z*:

(KD.2a) $\frac{\partial x}{\partial z} = \frac{1}{2}$ and $\frac{\partial x}{\partial z^*} = \frac{1}{2}$

(KD.2b) $\frac{\partial y}{\partial z} = -\frac{i}{2}$ and $\frac{\partial y}{\partial z^*} = +\frac{i}{2}$

From this we derive the derivatives with respect to z and z*:

(KD.3a) $\frac{\partial}{\partial z} = \frac{\partial x}{\partial z}\frac{\partial}{\partial x} + \frac{\partial y}{\partial z}\frac{\partial}{\partial y}$

$$= \frac{1}{2}\frac{\partial}{\partial x} - \frac{i}{2}\frac{\partial}{\partial y}$$

(KD.3b) $\frac{\partial}{\partial z^*} = \frac{\partial x}{\partial z^*}\frac{\partial}{\partial x} + \frac{\partial y}{\partial z^*}\frac{\partial}{\partial y}$

$$= \frac{1}{2}\frac{\partial}{\partial x} + \frac{i}{2}\frac{\partial}{\partial y}$$

Consider an arbitrary complex function:

(KD.4) $f(z) = u(x,y) + i\, v(x,y)$

Where both $u(x,y)$ **and** $v(x,y)$ are real functions.

We are now in the position to determine if this function contains z^*.

(KD.5) $\dfrac{\partial f}{\partial z^*} = \dfrac{1}{2}(\dfrac{\partial}{\partial x} + i\dfrac{\partial}{\partial y}) (u + i v)$

$$= \dfrac{1}{2}\left(\dfrac{\partial u}{\partial x} - \dfrac{\partial v}{\partial y}\right) + \dfrac{i}{2}(\dfrac{\partial v}{\partial x} + \dfrac{\partial u}{\partial y})$$

No dependence on z^* implies that both the real and imaginary parts of equation KD.5 are each equal to zero.

(KD.6a) $\dfrac{\partial u}{\partial x} = \dfrac{\partial v}{\partial y}$

(KD.6b) $\dfrac{\partial v}{\partial x} = -\dfrac{\partial u}{\partial y}$

These are known as the Cauchy-Riemann equations. They set the conditions for which a complex function is holomorphic.

KE. Complex Integration

From the fundamental theorem (equation EB.3), for $z = z(t)$, where $a \le t \le b$ we have,

(KE.1) $\int_a^b f(z)dz = \int_a^b u(t)dt + i \int_a^b v(t)dt$

Where the integral of the complex function $f(z)$ has been expressed into two integrals of the real functions $u(t)$ and $v(t)$. Each part is to be integrated separately.

KF. Integration with the use of the Path

Let C be a piecewise smooth path. Let $f(z)$ be a continuous function on C. For $z = z(t),$ where $a \le t \le b$, then we have

(KF.1) $\int_C f(z)dz = \int_a^b f[z(t)]\dot{z}\, dt$

Proof: The derivative $\dot{z} = \dfrac{dz}{dt}$. Substitute.

QED.

For a closed loop,

(KF.2) $\oint f(z)dz = 0$

Proof: $\oint f(z)dz = \int_a^b f(z)dz + \int_b^a f(z)dz$

$$= \int_a^b f(z)dz - \int_a^b f(z)dz = 0$$

QED.

KG. Basic Integral

Where C is a unit circle, then

(KG.1) $\oint \dfrac{dZ}{Z} = 2\pi i$

A unit circle can be represented by

(KG.2) $z(t) = e^{it}$
Then

(KG.3) $\frac{1}{z(t)} = e^{-it}$

And

(KG.4) $dz = ie^{it}dt$

Substitute equations KG.2-4 into KG.1,

(KG.5) $\oint \frac{dz}{z} = \int_0^{2\pi} e^{-it}ie^{it}dt = i\int_0^{2\pi} dt = 2\pi i$

KH. Cauchy's Integral Formula

Let $f(z)$ be a holomorphic function in a domain D. Then for any point z_0 in D and any simple closed path C in D that encloses z_0,

(KH.1) $\oint \frac{f(z)}{z-z_0} dz = 2\pi i f(z_0)$

Proof: $\oint \frac{f(z)}{z-z_0} dz = f(z_0) \oint \frac{dz}{z-z_0} + \oint \frac{f(z)-f(z_0)}{z-z_0} dz$

The first integral is $2\pi i$ (equation KG.1), and the second integral is zero (equation KF.2).

QED.

Note that we can conveniently express such function at a point z_0 as:

(KH.2) $f(z_0) = \frac{1}{2\pi i} \oint \frac{f(z)}{z-z_0} dz$

The function $f(z)$ is said to have a singularity at z_0.

KI. The Residue Theorem

Suppose we have a function such as:

(KI.1) $f(z) = \sum_{n=-\infty}^{n=\infty} a_n^j (z - z_j)^n$

If we take the complex integral along a curve C in the domain D where the function is holomorphic, then,

(KI.2) $\oint f(z)dz = \oint \sum_{n=-\infty}^{n=\infty} a_n^j (z - z_j)^n dz$

$= \sum_{n=-\infty}^{n=\infty} a_n^j \oint (z - z_j)^n dz$

$= 2\pi i\, a_{-1}^j$

The function $f(z)$ has singularity at $z = z_j$, for $n = -1$
The coefficient a_{-1}^j is called a residue. If the exponent of the residue is one, it is a simple residue. Otherwise, it is of order n.

Summing up over all the integrals for each singularity yields the Residue Theorem:

(KI.3) $\oint f(z)dz = 2\pi i \sum_{j=1}^{k} residues$

Residues are computed by finding the limit of the function as z approaches each singularity as follows:

(KI.4) $residue = \lim_{z \to z_0} \frac{1}{(k-1)!} \frac{d^{k-1}}{dz^{k-1}} [(z - z_0)^k f(z)]$

Example: Compute the integral of $\oint \frac{4z-9}{z(z-3)^2}\, dz$ over a circle of radius r=4, centered at the origin.

→ This has two singularities: a simple residue at $z = 0$; **and a residue of order 2 at $z = 3$.**

(a) simple residue at $z = 0$, that is, $z_0 = 0$, and k =1

$$\to \lim_{z \to z_0}(z - z_0)f(z) = \lim_{z \to 0}(z - 0)\frac{4z-9}{z(z-3)^2} = \lim_{z \to 0}\frac{4z-9}{(z-3)^2}$$

$$= \frac{-9}{(-3)^2} = -1$$

(b) a residue of order 2 at $z = 3$, that is, $z_0 = 3$, and k=2.

$$\to \lim_{z \to 3}\frac{d}{dz}[(z - 3)^2 \frac{4z-9}{z(z-3)^2}] = \lim_{z \to 3}\frac{d}{dz}[\frac{4z-9}{z}]$$

$$= \lim_{z \to 3}\frac{d}{dz}\left(4 - \frac{9}{z}\right)$$
$$= \lim_{z \to 3}\frac{d}{dz}\left(-\frac{9}{z}\right)$$
$$= \lim_{z \to 3}\frac{9}{z^2} = 1$$

Therefore,

$$\oint \frac{5z - 2}{z(z - 3)^2}\, dz = 2\pi i \sum_{j=1}^{2} residues$$
$$= 2\pi i(-1 + 1)$$
$$= 0$$

Index

Printed in the United States
By Bookmasters